Distortion and Stress

A Mechanical Designers' Workbook

Other Books in
The McGraw-Hill Mechanical Designers' Workbook Series

MACHINE DESIGN FUNDAMENTALS: A Mechanical Designers' Workbook

CORROSION AND WEAR: A Mechanical Designers' Workbook

FASTENING, JOINING, CONNECTING: A Mechanical Designers' Workbook

GEARING: A Mechanical Designers' Workbook

MECHANISMS: A Mechanical Designers' Workbook

BEARINGS AND LUBRICATION: A Mechanical Designers' Workbook

POWER TRANSMISSION ELEMENTS: A Mechanical Designers' Workbook

Distortion and Stress

A Mechanical Designers' Workbook

Editors in Chief

Joseph E. Shigley

Professor Emeritus
The University of Michigan
Ann Arbor, Michigan

Charles R. Mischke

Professor of Mechanical Engineering
Iowa State University
Ames, Iowa

McGraw-Hill Publishing Company

New York   St. Louis   San Francisco   Auckland   Bogotá
Caracas   Hamburg   Lisbon   London   Madrid   Mexico
Milan   Montreal   New Delhi   Oklahoma City
Paris   San Juan   São Paulo   Singapore
Sydney   Tokyo   Toronto

Library of Congress Cataloging-in-Publication Data

Main entry under title:

Standard handbook of machine design.

    Includes index.
    1. Machinery—Design—Handbooks, manuals, etc.
I. Shigley, Joseph Edward. II. Mischke, Charles R.
TJ230.S8235 1986   621.8′15   85-17079
ISBN 0-07-056892-8
ISBN 0-07-056924-X (workbook)

1234567890 KGP/KGP 8965432109

ISBN 0-07-056924-X

The material in this volume has been published previously in *Standard Handbook of Machine Design* by Joseph E. Shigley and Charles R. Mischke. Copyright © 1986 by McGraw-Hill, Inc. All rights reserved.

The editors for this book were Robert Hauserman and Scott Amerman, and the production supervisor was Dianne Walber. It was set in Times Roman by Techna Type.

Printed and bound by The Kingsport Press.

For more information about other McGraw-Hill materials, call 1-800-2-MCGRAW in the United States. In other countries, call your nearest McGraw-Hill office.

In Loving Memory of
Opal Shigley

CONTENTS

CONTRIBUTORS

Sachindranarayan Bhaduri, *Associate Professor,* Mechanical and Industrial Engineering Department, The University of Texas at El Paso, Texas

Harry Herman, *Professor of Mechanical Engineering,* New Jersey Institute of Technology, Newark, New Jersey

Charles R. Mischke, *Professor of Mechanical Engineering,* Iowa State University, Ames, Iowa

Joseph E. Shigley, *Emeritus Professor of Mechanical Engineering,* The University of Michigan, Ann Arbor, Michigan

PREFACE

There is no shortage of good textbooks treating the subject of machine design and related topics of study. But the beginning designer quickly learns that there is a great deal more to successful design than is presented in textbooks or taught in technical schools or colleges. A handbook connects formal education and the practice of design engineering by including the general knowledge required by every machine designer.

Much of the practicing designer's daily informational needs are satisfied in various pamphlets or brochures such as are published by the various standards organizations as well as manufacturers of various components used in design. Other sources include research papers, design magazines, and corporate publications concerned with specific products. More often than not, however, a visit to a design library or to a file cabinet will reveal that a specific publication is on loan, lost, or out of date. A handbook is intended to serve such needs quickly and immediately by giving the designer authoritative, up-to-date, understandable, and informative answers to the hundreds of such questions that arise every day in the work of a designer.

The *Standard Handbook of Machine Design** was written for working designers, and its place is on their desks, not on their bookshelves, for it contains a great many formulas, tables, charts, and graphs, many in condensed form. These are intended to give quick answers to the many questions that seem to arise constantly.

The *Mechanical Designers' Workbook* series consists of eight volumes, each containing a group of related topics selected from the *Standard Handbook of Machine Design.* Limiting each workbook to a single subject area of machine design makes it possible to create a thin, convenient volume bound in such a manner as to open flat and provide an opportunity to enter notes, references, graphs, equations, standard corporate practices, and other useful data. In fact, each chapter in every workbook contains gridded pages located in critical sections for this specific purpose. This flat-opening workbook is easier to use on the designer's work space and will save much wear and tear on the source handbook.

This workbook is devoted to fundamentals of distortion and stress utilized by any practicing mechanical designer. Chapter 1 tabulates common sections and their properties, preferred numbers and sizes, tolerances of steel sheets and bars, wire and sheet metal, and structural shapes. Chapter 2 has a concise summary of triaxial stress, stress-strain relationships, flexure, and stresses due to temperature and contact. Chapter 3 tabulates beam bending and also presents a deflection analysis method and computer-aided (BASIC) programming. Frame analysis is done using Castigliano's theorem.

---

*By Joseph E. Shigley and Charles R. Mischke, Coeditors-in-Chief, McGraw-Hill Publishing Company, New York, 1986.

Chapter 4 addresses instabilities in beams and columns, considers Euler's equation and its generalization, as well as discusses the column problem, including the beam-column problem. The chapter concludes with instabilities in beams. Chapter 5 applies Castigliano's theorem to rings and ring segments. Chapter 6 considers pressure cylinders of thin and thick wall varieties, and thermal stresses, and it concludes with comments on design. Chapter 7 develops the fundamental tolerance equations together with a scheme for analyzing assemblies with a number of pieces. An absolute tolerance worksheet and a statistical tolerance worksheet are developed. Because tolerance information is so scattered, extra pages are provided for notes.

Most of the artwork was competently prepared and supervised by Mr. Gary Roys of Madrid, Iowa, to whom the editors are indebted.

Care has been exercised to avoid error. The editors will appreciate being informed of errors discovered, so that they may be eliminated in subsequent printings.

JOSEPH E. SHIGLEY
CHARLES R. MISCHKE

chapter **1**
SECTIONS AND SHAPES— TABULAR DATA

JOSEPH E. SHIGLEY
Professor Emeritus
The University of Michigan
Ann Arbor, Michigan

1-1 CENTROIDS AND CENTER OF GRAVITY

When forces are distributed over a line, an area, or a volume, it is often necessary to determine where the resultant force of such a system acts. To have the same effect, the resultant must act at the centroid of the system. The *centroid* of a system is a point at which a system of distributed forces may be considered concentrated with exactly the same effect.

Figure 1-1 shows four weights W_1, W_2, W_4, and W_5 attached to a straight horizontal rod whose weight W_3 is shown acting at the center of the rod. The centroid of this *weight* or *point group* is located at G, which may also be called the *center of gravity* or the *center of mass* of the point group. The total weight of the group is

$$W = W_1 + W_2 + W_3 + W_4 + W_5$$

This weight, when multiplied by the *centroidal distance* $\bar{x}$ must balance or cancel the sum of the individual weights multiplied by their respective distances from the left end. In other words,

$$W\bar{x} = W_1 l_1 + W_2 l_2 + W_3 l_3 + W_4 l_4 + W_5 l_5$$

or

$$\bar{x} = \frac{W_1 l_1 + W_2 l_2 + W_3 l_3 + W_4 l_4 + W_5 l_5}{W_1 + W_2 + W_3 + W_4 + W_5}$$

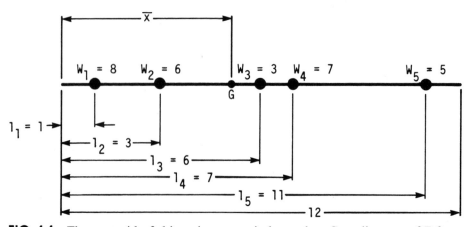

FIG. 1-1   The centroid of this point group is located at G, a distance of $\bar{x}$ from the left end.

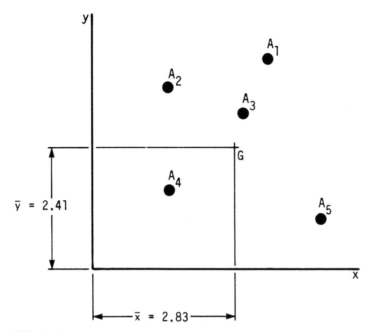

FIG. 1-2 The weightings and coordinates of the points are designated as A_i (x_i, y_i); they are $A_1 = 0.5(3.5, 4.0)$, $A_2 = 0.5(1.5, 3.5)$, $A_3 = 0.5(3.0, 3.0)$, $A_4 = 0.7(1.5, 1.5)$, and $A_5 = 0.7(4.5, 1.0)$.

A similar procedure can be used when the point groups are contained in an area such as Fig. 1-2. The centroid of the group at G is now defined by the two centroidal distances $\bar{x}$ and $\bar{y}$, as shown. Using the same procedure as before, we see that these must be given by the equations

$$\bar{x} = \sum_{i=1}^{i=N} A_i x_i \bigg/ \sum_{i=1}^{i=N} A_i \qquad \bar{y} = \sum_{i=1}^{i=N} A_i y_i \bigg/ \sum_{i=1}^{i=N} A_i \qquad (1\text{-}1)$$

A similar procedure is used to locate the centroids of a group of lines or a group of areas. Area groups are often composed of a combination of circles, rectangles, triangles, and other shapes. The areas and locations of the centroidal axes for many such shapes are listed in Table 1-1. For these, the x_i and y_i of Eqs. (1-1) are taken as the distances to the centroid of each area A_i.

Equations (1-1) can easily be solved on an ordinary calculator using the Σ key twice, once for the denominator and again for the numerator. The equations are also easy to program. Practice these techniques using the data and results in Figs. 1-1, 1-2, and 1-3.

By substituting integration signs for the summation signs in Eqs. (1-1), we get the more general form of the relations as

$$\bar{x} = \frac{\int x'\, dA}{\int dA} \qquad \bar{y} = \frac{\int y'\, dA}{\int dA}$$

These reduce to

$$\bar{x} = \frac{1}{A} \int x'\, dA \qquad \bar{y} = \frac{1}{A} \int y'\, dA \qquad (1\text{-}2)$$

(continued on p. 14)

TABLE 1-1 Properties of Sections†

1. Rectangle

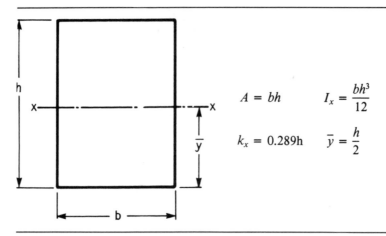

$$A = bh \qquad I_x = \frac{bh^3}{12}$$

$$k_x = 0.289h \qquad \bar{y} = \frac{h}{2}$$

2. Hollow rectangle

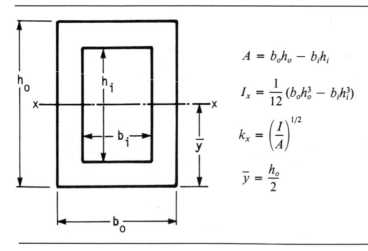

$$A = b_o h_o - b_i h_i$$

$$I_x = \frac{1}{12}(b_o h_o^3 - b_i h_i^3)$$

$$k_x = \left(\frac{I}{A}\right)^{1/2}$$

$$\bar{y} = \frac{h_o}{2}$$

3. Two rectangles

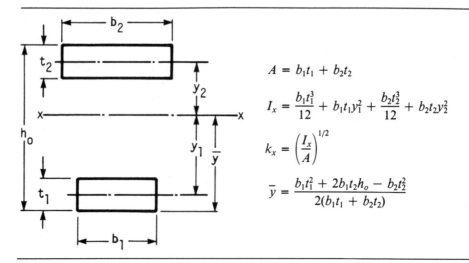

$$A = b_1 t_1 + b_2 t_2$$

$$I_x = \frac{b_1 t_1^3}{12} + b_1 t_1 y_1^2 + \frac{b_2 t_2^3}{12} + b_2 t_2 y_2^2$$

$$k_x = \left(\frac{I_x}{A}\right)^{1/2}$$

$$\bar{y} = \frac{b_1 t_1^2 + 2b_1 t_2 h_o - b_2 t_2^2}{2(b_1 t_1 + b_2 t_2)}$$

TABLE 1-1 Properties of Sections (*Continued*)

4. Triangle

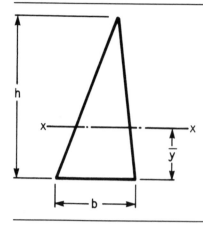

$$A = \frac{bh}{2} \qquad I_x = \frac{bh^3}{36}$$

$$k = 0.236h \qquad \bar{y} = \frac{h}{3}$$

5. Trapezoid

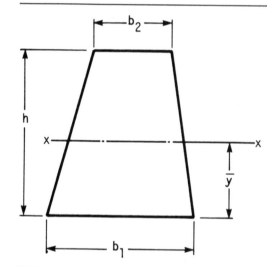

$$A = \frac{h}{2}(b_1 + b_2)$$

$$I_x = \frac{h^3(b_1^2 + 4b_1b_2 + b_2^2)}{36(b_1 + b_2)}$$

$$k_x = \frac{h[2(b_1^2 + 4b_1b_2 + b_2^2)]^{1/2}}{6(b_1 + b_2)}$$

$$\bar{y} = \frac{h(b_1 + 2b_2)}{3(b_1 + b_2)}$$

6. Circle

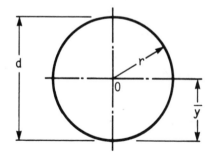

$$A = \pi r^2 = \frac{\pi d^2}{4}$$

$$I = \frac{\pi r^4}{4} = \frac{\pi d^4}{64}$$

$$k = \frac{r}{2} = \frac{d}{2} \qquad \bar{y} = r = \frac{d}{2}$$

†List of symbols: A = area; I = second area moment about principal axis; J_O = second polar area moment with respect to O; k = radius of gyration; and $\bar{x}, \bar{y}$ = centroidal distances.

TABLE 1-1 Properties of Sections (*Continued*)

7. Hollow circle

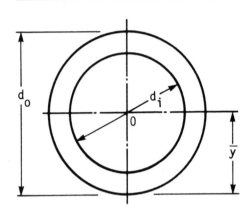

$$A = \frac{\pi}{4}(d_o^2 - d_i^2)$$

$$I = \frac{\pi}{64}(d_o^4 - d_i^4)$$

$$J_O = \frac{\pi}{32}(d_o^4 - d_i^4)$$

$$k = \frac{1}{4}(d_o^2 + d_i^2)^{1/2}$$

$$\bar{y} = \frac{d_o}{2}$$

8. Thin ring (annulus)

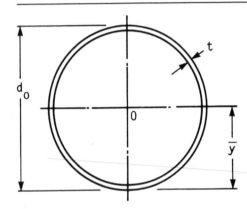

$$A = \pi d_o t \qquad I = \frac{\pi d_o^3 t}{8}$$

$$J_O = \frac{\pi d_o^3 t}{4}$$

$$k = 0.353 d_o \qquad \bar{y} = \frac{d_o}{2}$$

9. Semicircle

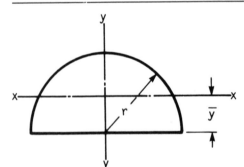

$$A = \frac{\pi r^2}{2} \qquad I_x = 0.1098 r^4$$

$$I_y = \frac{\pi r^4}{8} \qquad k_x = 0.264 r$$

$$k_y = \frac{r}{2} \qquad \bar{y} = 0.424 r$$

6

TABLE 1-1 Properties of Sections (*Continued*)

10. Circular sector

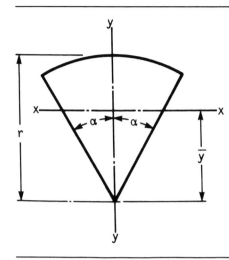

$$A = \alpha r^2$$

$$I_x = \frac{r^4}{4}\left(\alpha + \sin \alpha \cos \alpha - \frac{16}{9\alpha}\sin^2 \alpha\right)$$

$$I_y = \frac{r^4}{4}(\alpha - \sin \alpha \cos \alpha)$$

$$k_x = \frac{r}{2}\left(1 + \frac{\sin \alpha \cos \alpha}{\alpha} - \frac{16}{9\alpha}\sin^2 \alpha\right)^{1/2}$$

$$k_y = \frac{r}{2}\left(\frac{\alpha - \sin \alpha \cos \alpha}{\alpha}\right)^{1/2}$$

$$\bar{y} = \frac{2r \sin \alpha}{3\alpha}$$

11. Circular segment

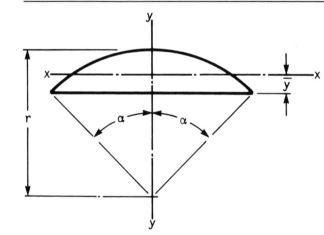

$$A = \frac{r^2}{2}(2\alpha - \sin 2\alpha)$$

$$I_x = r^4\left[\left(\frac{2\alpha - \sin 2\alpha}{8}\right)\left(1 + \frac{2\sin^3 \alpha \cos \alpha}{\alpha - \sin \alpha \cos \alpha}\right) - \frac{8\sin^6 \alpha}{9(2\alpha - \sin 2\alpha)}\right]$$

$$k_x = \frac{r}{2}\left[1 + \frac{2\sin^3 \alpha \cos \alpha}{\alpha - \sin \alpha \cos \alpha} - \frac{64\sin^6 \alpha}{9(2\alpha - \sin 2\alpha)^2}\right]^{1/2}$$

$$\bar{y} = \frac{4r \sin^3 \alpha}{6\alpha - 3\sin 2\alpha} - r \cos \alpha$$

TABLE 1-1 Properties of Sections (*Continued*)

12. Parabola

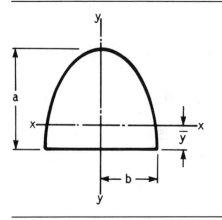

$$A = \frac{4ab}{3} \qquad I_x = \frac{16a^3b}{175}$$

$$I_y = \frac{4ab^3}{15} \qquad \bar{y} = \frac{a}{5}$$

13. Semiparabola

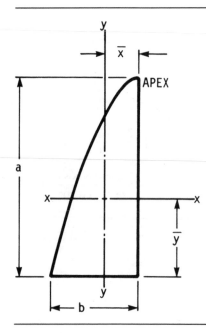

$$A = \frac{2ab}{3} \qquad I_x = \frac{8a^3b}{175}$$

$$I_y = \frac{19ab^3}{480} \qquad \bar{y} = \frac{2a}{5} \qquad \bar{x} = \frac{3b}{8}$$

TABLE 1-1 Properties of Sections (*Continued*)

14. Ellipse

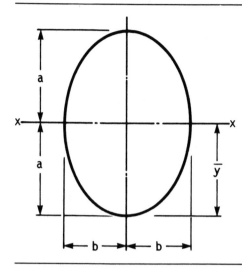

$$A = \pi ab \qquad I_x = \frac{\pi a^3 b}{4}$$

$$k_x = \frac{a}{2} \qquad \bar{y} = a$$

15. Semiellipse

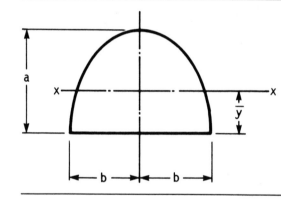

$$A = \frac{\pi ab}{2} \qquad\qquad I_x = a_3 b\left(\frac{\pi}{8} - \frac{8}{9\pi}\right)$$

$$k_x = \frac{b}{6\pi}(9\pi^2 - 64)^{1/2} \qquad \bar{y} = \frac{4a}{3\pi}$$

16. Hollow ellipse

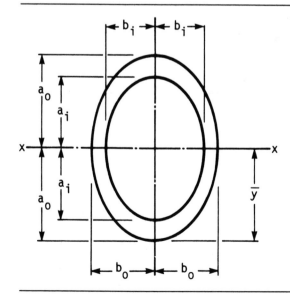

$$A = \pi(a_o b_o - a_i b_i)$$

$$I_x = \frac{\pi(a_o^3 b_o - a_i^3 b_i)}{4}$$

$$k_x = \frac{1}{2}\left(\frac{a_o^3 b_o - a_i^3 b_i}{a_o b_o - a_i b_i}\right)^{1/2} \qquad \bar{y} = a_o$$

TABLE 1-1 Properties of Sections (*Continued*)

17. Regular polygon (*N* sides)

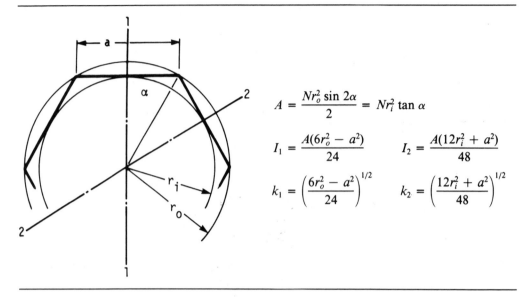

$$A = \frac{Nr_o^2 \sin 2\alpha}{2} = Nr_i^2 \tan \alpha$$

$$I_1 = \frac{A(6r_o^2 - a^2)}{24} \qquad I_2 = \frac{A(12r_i^2 + a^2)}{48}$$

$$k_1 = \left(\frac{6r_o^2 - a^2}{24}\right)^{1/2} \qquad k_2 = \left(\frac{12r_i^2 + a^2}{48}\right)^{1/2}$$

18. Angle

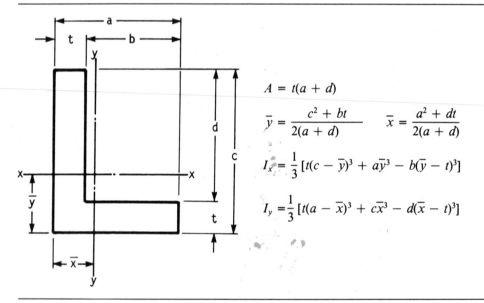

$$A = t(a + d)$$

$$\bar{y} = \frac{c^2 + bt}{2(a + d)} \qquad \bar{x} = \frac{a^2 + dt}{2(a + d)}$$

$$I_x = \frac{1}{3}[t(c - \bar{y})^3 + a\bar{y}^3 - b(\bar{y} - t)^3]$$

$$I_y = \frac{1}{3}[t(a - \bar{x})^3 + c\bar{x}^3 - d(\bar{x} - t)^3]$$

TABLE 1-1 Properties of Sections (*Continued*)

19. T section

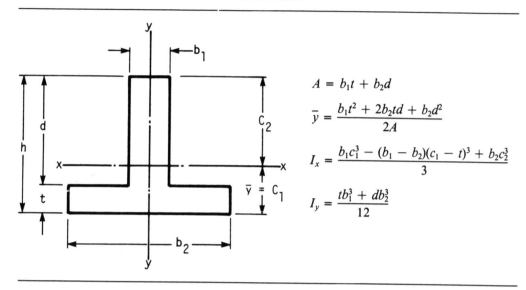

$$A = b_1 t + b_2 d$$

$$\bar{y} = \frac{b_1 t^2 + 2b_2 td + b_2 d^2}{2A}$$

$$I_x = \frac{b_1 c_1^3 - (b_1 - b_2)(c_1 - t)^3 + b_2 c_2^3}{3}$$

$$I_y = \frac{tb_1^3 + db_2^3}{12}$$

20. U Section

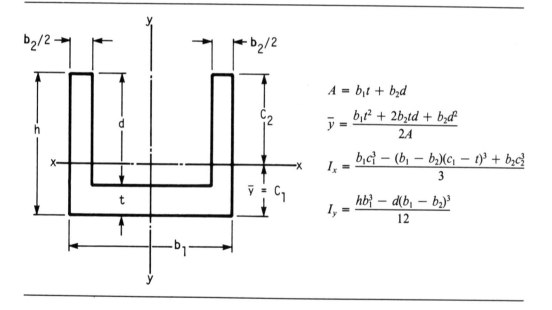

$$A = b_1 t + b_2 d$$

$$\bar{y} = \frac{b_1 t^2 + 2b_2 td + b_2 d^2}{2A}$$

$$I_x = \frac{b_1 c_1^3 - (b_1 - b_2)(c_1 - t)^3 + b_2 c_2^3}{3}$$

$$I_y = \frac{hb_1^3 - d(b_1 - b_2)^3}{12}$$

Notes ▪ Drawings ▪ Ideas

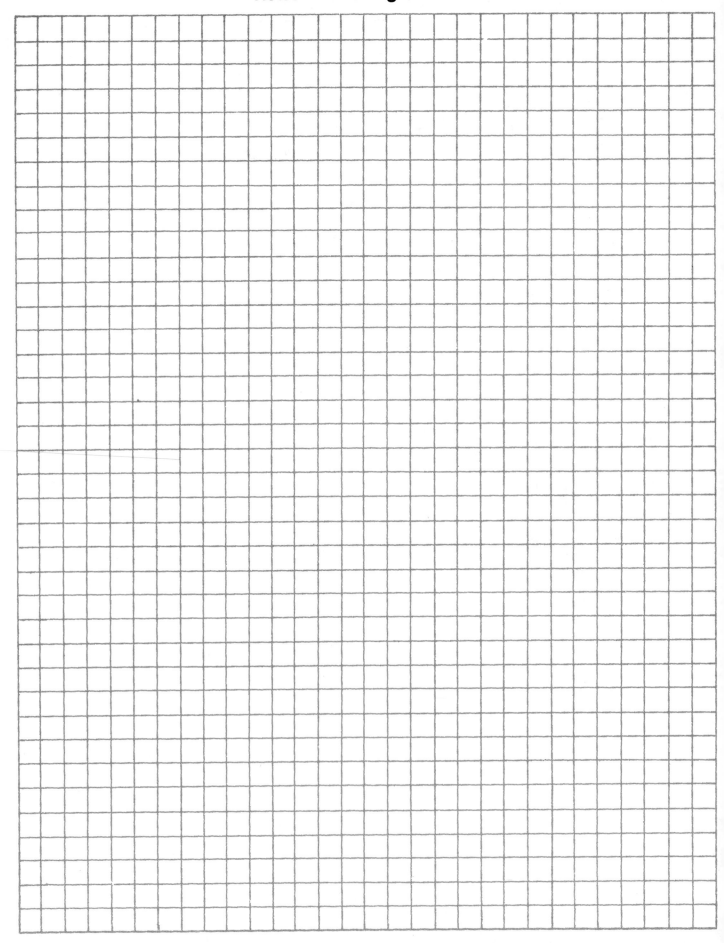

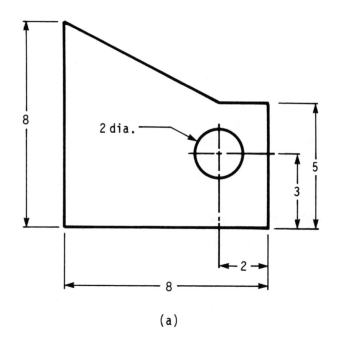

(a)

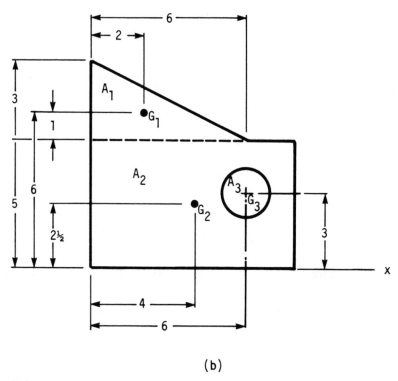

(b)

FIG. 1-3 A composite shape consisting of a rectangle, a triangle, and a circular hole. The centroidal distances are found to be $\bar{x} = 3.47$ and $\bar{y} = 3.15$.

where x' and y' = coordinate distances to the centroid of the element dA. These equations can be solved by

- Finding expressions for x' and y' and then performing the integration analytically.
- Approximate integration using the routines described in the programming manual of your programmable calculator or computer.
- Using numerical integration routines.

1-2   SECOND MOMENTS OF AREAS

The expression $Ax = \int x' \, dA$ from Eqs. (1-2) is a *first moment of an area*. A *second moment of an area* is obtained when the element of area is multiplied by the square of a distance to some stated axis. Thus the expressions

$$\int x^2 \, dA \qquad \int y^2 \, dA \qquad \int r^2 \, dA \quad (1\text{-}3)$$

are all second moments of areas. Such formulas resemble the equation for *moment of inertia*, which is

$$\int \rho^2 \, dm \qquad (1\text{-}4)$$

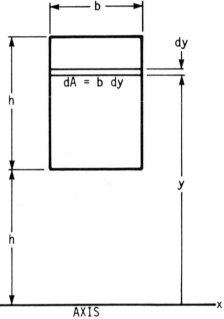

where ρ = distance to some axis, and dm = an element of mass. Because of the resemblance, Eqs. (1-3) are often called the equations for moment of inertia too, but this is a misnomer because an area cannot have inertia.

We can find the second moment of an area about rectangular axes by using one of the formulas

$$I_x = \int y^2 \, dA \qquad I_y = \int x^2 \, dA \quad (1\text{-}5)$$

FIG. 1-4   Moment of inertia of a rectangle.

EXAMPLE 1. Find the second moment of area of the rectangle in Fig. 1-4 about the x axis.

Solution. Select an element of area dA such that it is everywhere y units from x. Substituting appropriate terms into Eqs. (1-5) gives

$$I_x = \int y^2 \, dA = \int_h^{2h} y^2 b \, dy = \frac{by^3}{3}\Big|_h^{2h} = \frac{7bh^3}{3} \qquad Ans.$$

The *polar second moment of an area* is the second moment taken about an axis *normal to the plane* of an area. The equation is

$$J = \int \rho^2 \, dA \qquad (1\text{-}6)$$

EXAMPLE 2. Find the polar second moment of the area of a circle about its centroidal axis.

Solution. Let the radius of the circle be r. Define a thin elemental ring of thickness $d\rho$ at radius ρ. Then $dA = 2\pi\rho \, d\rho$. We now have

$$J = \int \rho^2 \, dA = \int_0^r \rho^2 (2\pi\rho) \, d\rho = 2\pi \left.\frac{\rho^4}{4}\right|_0^r = \frac{\pi r^4}{2} \qquad Ans.$$

1-2-1 Radius of Gyration

If we think of the second moment of an area as the total area times the square of a fictitious distance, then

$$I_x = \int y^2 \, dA = k_x^2 A \qquad \text{or} \qquad k_x = \sqrt{\frac{I_x}{A}} \tag{1-7}$$

In polar form,

$$J_z = \int \rho^2 \, dA = k_z^2 A \qquad \text{or} \qquad k_z = \sqrt{\frac{J_z}{A}} \tag{1-8}$$

In each case, k is called the *radius of gyration.*

1-2-2 Transfer Formula

In Fig. 1-5 suppose we know the second moment of the area about x to be I_x. We can find the second moment of the area about some new axis that is parallel to the old using the transfer formula. Thus the second moment of the area in Fig. 1-5 about the x' axis is

$$I' = I_G + d^2 A \tag{1-9}$$

where I_G = second moment about the centroidal axis, and d = transfer distance. Using this formula and the second moments from Table 1-1 makes it possible to compute the second moments of sections made up of a combination of shapes. The procedure has much in common with the example in Fig. 1-3.

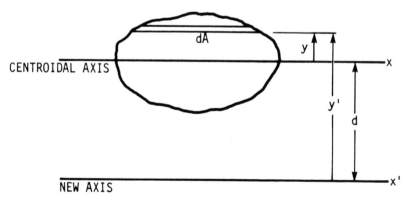

FIG. 1-5 Use of the transfer formula.

Notes ▪ Drawings ▪ Ideas

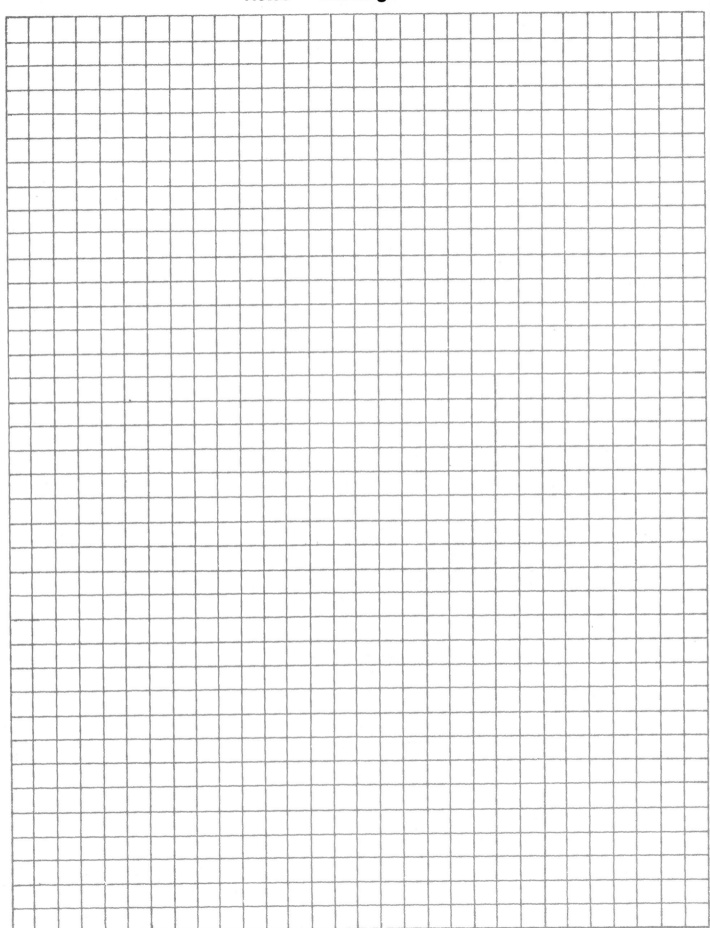

1-2-3 Principal Axes

Sometimes we encounter the integral

$$I_{xy} = \int xy \, dA \tag{1-10}$$

which is called the *product moment of an area.* This integral can be either positive or negative because x and y can have positive or negative values.

If one of the axes of an area, say y, is an axis of symmetry, then every element of area dA located by a positive x will have a twin, symmetrically located, having a corresponding negative x. These will sum to zero in the integration, and so the product moment is always zero when either x or y is an axis of symmetry. Since I_{xy} can be either positive or negative, there must be some orientation of rectangular axes where $I_{xy} = 0$. The two axes corresponding to this zero position are called the *principal axes.* If such axes intersect at the centroid of a section, then they are called the *centroidal principal axes.*

1-3 PREFERRED NUMBERS AND SIZES

The recommendations given in this section are not intended to be used as rules for design since there are none. And even if rules were specified, there would be many occasions when designers would have to deviate from them, because other more pressing considerations may be present.

1-3-1 Preferred Numbers

A set of characteristic values that are to be distributed over a specified range for machines or products can be best obtained using a set of *preferred numbers.* Examples are the horsepower ratings of electric motors, the capacities of presses, or the speeds of a truck transmission. The preferred number system is internationally standardized (ISO3) and is described as the Renard, or R, series. This series is shown in Table 1-2. Some of the interesting characteristics are

1. The series can be applied to any value because it can be increased or decreased by powers of 10.
2. The R10 series contains all the R5 values; the R20 series contains all the R10 values; etc.
3. The preferred numbers can be multiplied, divided, or raised to a power by a single number, and the result, even if slightly rounded, is still a preferred number.
4. The number $3.15 \cong \pi$ in R10 and up means that the diameter, area, and circumference of a circle are also preferred numbers in view of the previous characteristic.

Preferred numbers are based on logarithmic interpolation and are given by

$$x_i = x_s \left(\frac{x_l}{x_s} \right)^{i/n+1} \tag{1-11}$$

where x_l = largest number
x_s = smallest number

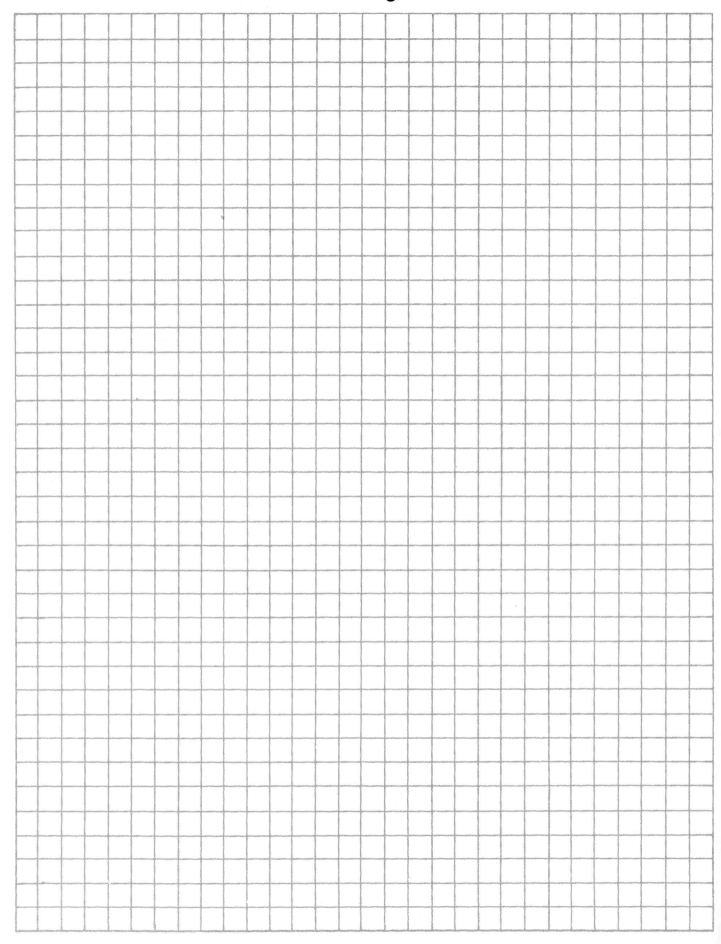

TABLE 1-2 Preferred Numbers

First choice R5	Second choice R10	Third choice R20	Fourth choice R40
1			
		1.12	1.06
	1.25		1.18
		1.4	1.32
			1.5
1.6		1.8	1.7
	2		1.9
		2.24	2.12
			2.36
2.5		2.8	2.65
	3.15		3
		3.55	3.35
			3.75
4		4.5	4.25
	5		4.75
		5.6	5.3
			6
6.3		7.1	6.7
	8		7.5
		9	8.5
			9.5
10			

SOURCE: British standard PD 6481-1977.

$x_i = i$th interpolated number
$i = $ interpolation number
$n = $ number of interpolations

The results should be appropriately rounded.

1-3-2 Preferred Sizes

Table 1-3 provides a list of preferred sizes for linear measurement in SI units. Note that these sizes are not the same as preferred numbers because of the convenience and simplicity of whole numbers for sizes of things. Preferred sizes in fractions of inches are listed in Table 1-4.

1-4 SIZES AND TOLERANCES OF STEEL SHEETS AND BARS

The dimensions and tolerances of steel products in this section are given in U.S. Customary System (USCS) units. Multiply inches by 25.4 to get the units of millimeters.

1-4-1 Sheet Steel

The *Manufacturer's Standard Gauge* for iron and steel sheets specifies a gauge number based on the weight per square foot. Remember, the gauge size is based on

TABLE 1-3 Preferred Metric Sizes in Millimeters

1st choice	2nd choice	3rd choice	1st choice	2nd choice	3rd choice	1st choice	2nd choice	3rd choice	1st choice	2nd choice	3rd choice
1	1.1		10			100		102	200		205
1.2			12	11		105		108		210	215
	1.4	1.3			13	110		112	220		225
		1.5		14	15		115	118		230	235
1.6		1.7	16		17	120		122	240		245
	1.8	1.9		18	19		125	128		250	255
2		2.1	20		21	130		132	260		265
	2.2	2.4		22	23		135	138		270	275
2.5		2.6			24	140		142	280		285
	2.8		25		26		145	148		290	295
3		3.2		28		150		152	300		305
	3.5	3.8	30	32			155	158		310	315
4		4.2			34	160		162	320		325
	4.5	4.8	35		36		165	168		330	335
5		5.2	40	38		170		172	340		345
	5.5	5.8		42			175	178		350	355
6			45		44	180		182	360		365
	6.5	6.8		48	46		185	188		370	375
	7	7.5	50	52		190		192	380		385
8		8.5	55		54		195	198		390	395
10	9	9.5	60	58	56	200			400†		
			65	62	64						
			70	68	66						
			75	72	74						
			80	78	76						
			90	85	82						
			100	95	88						
					92						
					98						

†Continued similarly above 400 mm.
SOURCE: British standard PD 6481-1977.

TABLE 1-4 Preferred Sizes in Fractions of Inches†

$\frac{1}{64}$	$\frac{1}{2}$	$2\frac{1}{4}$	5	$9\frac{1}{2}$	15
$\frac{1}{32}$	$\frac{9}{16}$	$2\frac{1}{2}$	$5\frac{1}{4}$	10	$15\frac{1}{2}$
$\frac{1}{16}$	$\frac{5}{8}$	$2\frac{3}{4}$	$5\frac{1}{2}$	$10\frac{1}{2}$	16
$\frac{3}{32}$	$\frac{11}{16}$	3	$5\frac{3}{4}$	11	$16\frac{1}{2}$
$\frac{1}{8}$	$\frac{3}{4}$	$3\frac{1}{4}$	6	$11\frac{1}{2}$	17
$\frac{5}{32}$	$\frac{7}{8}$	$3\frac{1}{2}$	$6\frac{1}{2}$	12	$17\frac{1}{2}$
$\frac{3}{16}$	1	$3\frac{3}{4}$	7	$12\frac{1}{2}$	18
$\frac{1}{4}$	$1\frac{1}{4}$	4	$7\frac{1}{2}$	13	$18\frac{1}{2}$
$\frac{5}{16}$	$1\frac{1}{2}$	$4\frac{1}{4}$	8	$13\frac{1}{2}$	19
$\frac{3}{8}$	$1\frac{3}{4}$	$4\frac{1}{2}$	$8\frac{1}{2}$	14	$19\frac{1}{2}$
$\frac{7}{16}$	2	$4\frac{3}{4}$	9	$14\frac{1}{2}$	20

†See also ANSI standard Z17.1-1973, Preferred Numbers.

TABLE 1-5 Gauge Sizes of Carbon Steel Sheets

Gauge no.	Thickness, in†	Weight, lb/ft²‡	Gauge no.	Thickness, in†	Weight, lb/ft²‡
7	0.1793	7.500	23	0.0269	1.125
8	0.1644	6.875	24	0.0239	1.000
9	0.1494	6.250	25	0.0209	0.875
10	0.1345	5.625	26	0.0179	0.750
11	0.1196	5.000	27	0.0164	0.6875
12	0.1046	4.375	28	0.0149	0.625
13	0.0897	3.750	29	0.0135	0.5625
14	0.0747	3.125	30	0.0120	0.500
15	0.0673	2.812	31	0.0105	0.4375
16	0.0598	2.500	32	0.0097	0.4062
17	0.0538	2.250	33	0.0090	0.375
18	0.0478	2.000	34	0.0082	0.3438
19	0.0418	1.750	35	0.0075	0.3125
20	0.0359	1.500	36	0.0067	0.2812
21	0.0329	1.375	37	0.0064	0.2656
22	0.0299	1.250	38	0.0060	0.250

†Multiply the thickness in inches by 25.4 to get the thickness in millimeters.

‡Multiply the weight in pounds per square foot by 4.88 to get the mass in kilograms per square meter (SI units).

weight, not thickness. Steel products having thicknesses of $\frac{1}{4}$ in and over are called *plates* or *flats,* depending on the width.

The weights and equivalent thicknesses of carbon steel sheets are shown in Table 1-5. Standard widths and lengths available depend on the gauge sizes. Most are also available in coils, but not all sizes may be stocked by steel warehouses.

Tables 1-6 to 1-11 provide the thickness tolerances for various grades of steel sheets. Except as noted, the tables apply to both coils and cut lengths. The width ranges are from *over* the lower limit up to and *including* the upper limit.

TABLE 1-6 Thickness Tolerances for Hot-Rolled Carbon Sheets†

Thickness, in	Width, in					
	12–20	20–40	40–48	48–60	60–72	72 up
0.0449–0.0508	5	5	5			
0.0509–0.0567	5	5	6	6	7	
0.0568–0.0709	6	6	6	7	7	
0.0710–0.0971	6	7	7	7	8	8
0.0972–0.1799	7	7	8	8	8	8
0.1800–0.2299	7	8	9			

†Tolerances are plus or minus and in mils (1 mil = 0.001 in). This table applies only to coils.

SOURCE: Ref. [1-1], Sec. 5, Aug. 1979.

TABLE 1-7 Thickness Tolerances for Hot-Rolled Alloy Steel Sheets†

Thickness, in	Width, in					
	24–32	32–40	40–48	48–60	60–72	72–80
0.0568–0.0709	6	6	6	7	7	
0.0710–0.0821	7	7	7	7	8	8
0.0822–0.0971	7	8	8	8	9	9
0.0972–0.1799	8	9	10	10	11	12
0.1800–0.2299	9	9	10			

†Tolerances are plus or minus and in mils (1 mil = 0.001 in).
SOURCE: Ref [1-1], Sec. 5, Aug. 1979.

1-4-2 Bar Steel

When hot-rolled bars are machined on centers it is necessary to allow for straightness as well as for the size and out-of-round tolerances in selecting the diameter (see Table 1-12). Tolerances for cold-finished bars are given in Tables 1-13, 1-14, and 1-15.

1-4-3 Pipe and Tubing

The outside diameter of pipe having a nominal size of 12 in or smaller is larger than the nominal size. The difference between pipe and tubing is that pipe is intended to be used in piping systems; also, tubing has an outside diameter the same as the nominal size. See Table 1-16 for pipe sizes.

Seamless mechanical steel tubing is available in a great range of sizes from about $\frac{3}{16}$ in outside diameter with a wall thickness of no. 24 gauge B and W up to a wall thickness of 1 in and an outside diameter of 12 in or over. Welded tubing is made from strip steel, either hot rolled with a bright finish or cold rolled. Tubing is also available cold drawn and may be obtained with a high-quality inside finish for certain applications.

The wall thickness of tubing is usually specified in gauge sizes or in fractions of an inch when USC units are used. The tolerances of tubing are generally specified for the outside diameter and the wall thickness. This means that the inside diameter

(continued on p. 31)

TABLE 1-8 Thickness Tolerances for Hot-Rolled High-Strength Steel Sheets†

Thickness, in	Width, in					
	12–15	15–20	20–32	32–40	40–48	48–60
0.0710–0.0821	6	7	7	7	7	7
0.0822–0.0971	6	7	7	8	8	8
0.0972–0.1799	7	8	8	9	10	10
0.1800–0.2299	7	8	9	9	10	

†Tolerances are plus or minus and in mils (1 mil = 0.001 in).
SOURCE: Ref. [1-1], Sec. 5, Aug. 1979.

TABLE 1-9 Thickness Tolerances for Cold-Rolled Carbon Steel Sheets†

Thickness, in	Width, in			
	2–12	12–15	15–72	72 up
0.0142–0.0194	2	2	2	
0.0195–0.0388	3	3	3	3
0.0389–0.0567	4	4	4	4
0.0568–0.0709	5	5	5	5
0.0710–0.0971	5	5	5	6
0.0972–0.1419	. . . .	5	6	7

†Tolerances are plus or minus and in mils (1 mil = 0.001 in).
SOURCE: Ref [1-1], Sec. 5, Aug. 1979.

TABLE 1-10 Thickness Tolerances for Cold-Rolled Alloy Steel Sheets†

Thickness, in	Width, in					
	24–32	32–40	40–48	48–60	60–70	70–80
0.0195–0.0313	3	3	3	3		
0.0314–0.0508	4	4	4	4	5	
0.0509–0.0567	5	5	5	5	6	
0.0568–0.0709	5	5	5	6	6	
0.0710–0.0821	5	6	6	6	7	7
0.0822–0.0971	6	7	7	8	9	9
0.0972–0.1419	7	8	9	10	10	11
0.1420–0.1799	8	9	10	10	11	12
0.1800–0.2299	8	9	10			

†Tolerances are plus or minus and in mils (1 mil = 0.001 in).
SOURCE: Ref. [1-1], Sec. 5, Aug. 1979.

TABLE 1-11 Thickness Tolerances for Cold-Rolled High-Strength Steel Sheets†

Thickness, in	Width, in					
	2–12	12–15	15–24	24–32	32–40	40–48
0.0142–0.0194	2	2	2	2	2	2
0.0195–0.0388	3	3	3	3	3	3
0.0389–0.0567	4	4	4	4	4	4
0.0568–0.0709	5	5	5	5	5	5
0.0710–0.0971	6	5	5	5	6	6
0.0972–0.1419	. . . .	5	6	6	6	6
0.1419 up	. . . .	6	6	7	7	7

†Tolerances are plus or minus and in mils (1 mil = 0.001 in).
SOURCE: Ref. [1-1], Sec. 5, Aug. 1979.

Notes · Drawings · Ideas

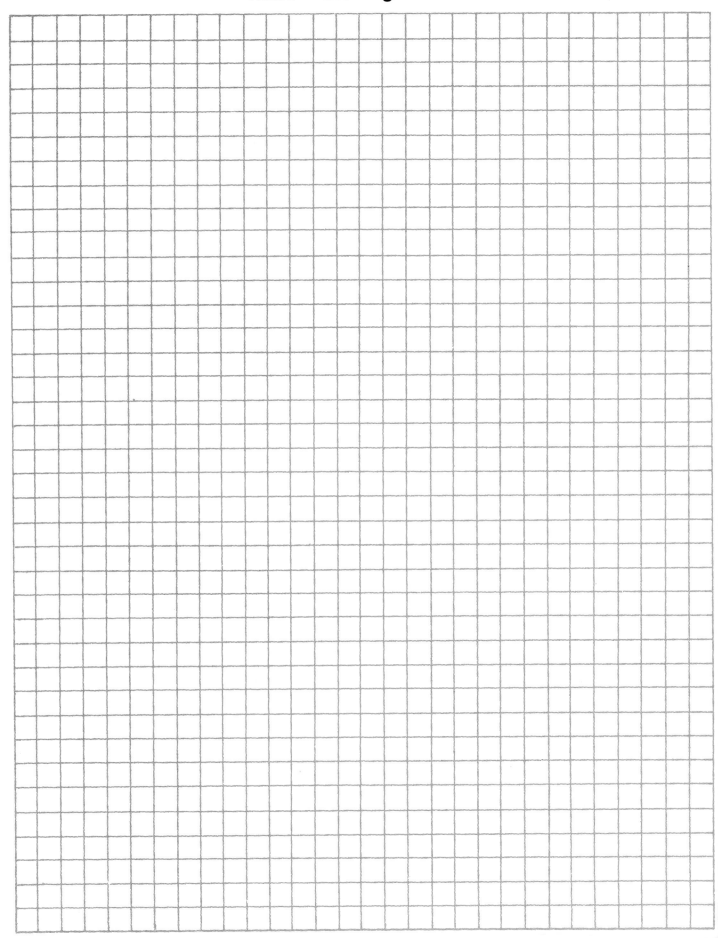

TABLE 1-12 Machining Allowances for Hot-Rolled Carbon Steel Bars for Turning on Centers†

Diameter, in	Allowance, in	Diameter, in	Allowance, in
To $\frac{7}{8}$	0.025	$2\frac{1}{2}$ to $3\frac{1}{2}$	0.090
$\frac{7}{8}$ to 1	0.028	$3\frac{1}{2}$ to $4\frac{1}{2}$	0.115
1 to $1\frac{1}{8}$	0.031	$4\frac{1}{2}$ to $5\frac{1}{2}$	0.140
$1\frac{1}{8}$ to $1\frac{1}{4}$	0.034	$5\frac{1}{2}$ to $6\frac{1}{2}$	0.165
$1\frac{1}{4}$ to $1\frac{3}{8}$	0.037	$6\frac{1}{2}$ to $8\frac{1}{4}$	0.209
$1\frac{3}{8}$ to $1\frac{1}{2}$	0.040	$8\frac{1}{4}$ to $9\frac{1}{2}$	0.240
$1\frac{1}{2}$ to 2	0.053	$9\frac{1}{2}$ to 10	0.253
2 to $2\frac{1}{2}$	0.065		

†Size range is from over the lower limit up to and including the upper limit; the allowances are on the *radius*.

SOURCE: Ref. [1-1].

TABLE 1-13 Size Tolerances for Cold-Drawn Carbon Steel Bars†

Size and shape, in	Carbon range, percent			
	To 0.28	0.28–0.55	To 0.55‡	Over 0.55§
Rounds:				
To $1\frac{1}{2}$	2	3	4	5
$1\frac{1}{2}$ to $2\frac{1}{2}$	3	4	5	6
$2\frac{1}{2}$ to 4	4	5	6	7
Hexagons:				
To $\frac{3}{4}$	2	3	4	6
$\frac{3}{4}$ to $1\frac{1}{2}$	3	4	5	7
$1\frac{1}{2}$ to $2\frac{1}{2}$	4	5	6	8
$2\frac{1}{2}$ to $3\frac{1}{8}$	5	6	7	9
Squares:				
To $\frac{3}{4}$	2	4	5	7
$\frac{3}{4}$ to $1\frac{1}{2}$	3	5	6	8
$1\frac{1}{2}$ to $2\frac{1}{2}$	4	6	7	9
$2\frac{1}{2}$ to 4	6	8	9	11
Flats				
To $\frac{3}{4}$	3	4	6	8
$\frac{3}{4}$ to $1\frac{1}{2}$	4	5	8	10
$1\frac{1}{2}$ to 3	5	6	10	12
3 to 4	5	6	10	12
4 to 6	8	10	12	20
Over 6	13	15		

†Includes tolerances for bars that have been annealed, spheroidize annealed, normalized, normalized and tempered, or quenched and tempered before cold finishing. The table *does not* include tolerances for bars that are spheroidize annealed, normalized, normalized and tempered, or quenched and tempered after cold finishing. Size range and carbon range are from over the lower limit up to and including the upper limit. Tolerances are minus and are in mils (1 mil = 0.001 in).

‡Stress relieved or annealed after cold finishing.

§Quenched and tempered or normalized and tempered before cold finishing.

¶These tolerances apply to *both* the widths and thickness of flats.

SOURCE: Ref. [1-1].

TABLE 1-14 Size Tolerances for Cold-Finished, Turned, and Polished Carbon Steel Round Bars†

Diameter, in	Carbon range, percent			
	To 0.28	0.28–0.55	To 0.55‡	Over 0.55§
To 1½	2	3	4	5
1½ to 2½	3	4	5	6
2½ to 4	4	5	6	7
4 to 6	5	6	7	8
6 to 8	6	7	8	9
8 to 9	7	8	9	10
Over 9	8	9	10	11

†Includes tolerances for bars that have been annealed, spheroidize annealed, normalized, normalized and tempered, or quenched and tempered before cold finishing. The table *does not* include tolerances for bars that are spheroidize annealed, normalized, normalized and tempered, or quenched and tempered after cold finishing. Size range and carbon range are from over the lower limit up to and including the upper limit. Tolerances are minus and are in mils (1 mil = 0.001 in).
‡Stress relieved or annealed after cold finishing.
§Quenched and tempered or normalized and tempered before cold finishing.
SOURCE: Ref. [1-1].

TABLE 1-15 Size Tolerances for Ground and Polished Carbon Steel Rounds Prefinished by Cold Drawing or by Turning†

Diameter, in	Prefinish	
	Cold drawn	Turned
To 1½	1	1
1½ to 2½	1.5	1.5
2½ to 3	2	2
3 to 4	3	3
4 to 6		4‡
Over 6		5‡

†Size range is from over the lower limit up to and including the upper limit. Tolerances are minus and in mils (1 mil = 0.001 in).
‡Increase this tolerance by 1 mil if the steels have a sulfur content under 0.08 percent or if they are thermally treated.
SOURCE: Ref. [1-1].

TABLE 1-16 Dimensions and Weights for Threaded and Coupled Pipe

Nominal size, in	Outside diameter, in	Wall thickness, in	Weight,† lb/ft	Weight class	Schedule no.
⅛	0.405	0.068	0.24	STD	40
		0.095	0.32	XS	80
¼	0.540	0.088	0.42	STD	40
		0.119	0.54	XS	80
⅜	0.675	0.091	0.57	STD	40
		0.126	0.74	XS	80

Nominal size, in	Outside diameter, in	Wall thickness, in	Weight,† lb/ft	Weight class	Schedule no.
½	0.840	0.109	0.85	STD	40
		0.147	1.09	XS	80
		0.294	1.72	XXS	
¾	1.050	0.113	1.13	STD	40
		0.154	1.48	XS	80
		0.308	2.44	XXS	
1	1.315	0.133	1.68	STD	40
		0.179	2.18	XS	80
		0.358	3.66	XXS	
1¼	1.660	0.140	2.28	STD	40
		0.191	3.02	XS	80
		0.382	5.22	XXS	
1½	1.900	0.145	2.73	STD	40
		0.200	3.66	XS	80
		0.400	6.41	XXS	
2	2.375	0.154	3.68	STD	40
		0.218	5.07	XS	80
		0.436	9.03	XXS	
2½	2.875	0.203	5.82	STD	40
		0.276	7.73	XS	80
		0.552	13.70	XXS	
3	3.500	0.216	7.62	STD	40
		0.300	10.33	XS	80
		0.600	18.57	XXS	
3½	4.000	0.226	9.20	STD	40
		0.318	12.63	XS	80
4	4.500	0.237	10.89	STD	40
		0.337	15.17	XS	80
		0.674	27.58	XXS	
5	5.563	0.258	14.81	STD	40
		0.375	21.09	XS	80
		0.750	38.61	XXS	
6	6.625	0.280	19.18	STD	40
		0.432	28.89	XS	80
		0.864	53.14	XXS	
8	8.625	0.277	25.55		30
		0.322	29.35	STD	40
		0.500	43.90	XS	80
		0.875	72.44	XXS	
10	10.750	0.279	32.75		
		0.307	35.75		30
		0.365	41.85	STD	40
		0.500	55.82	XS	60
12	12.750	0.330	45.45		30
		0.375	51.15	STD	
		0.500	66.71	XS	

†This is the weight of threaded pipe including the coupling.

SOURCE: ASTM standard A53, Table X3. A greater range of sizes together with SI equivalents is given in ANSI standard B36.10-1979.

TABLE 1-17 Decimal Equivalents of Wire and Sheet-Metal Gauges in Inches
Always specify the name of the gauge when gauge numbers are used.

Gauge Number	American or Brown & Sharpe (nonferrous sheet and rod)	Birmingham or Stubs iron wire (tubing, ferrous strip, flat wire, and spring steel)	United States Standard (ferrous sheet and plate, 480 lb/ft)	Manufacturers Standard (ferrous sheet)	Steel wire or Washburn & Moen (ferrous wire except music wire)	Music wire (music wire)	Stubs steel wire (steel drill rod)	Twist drill (twist drills and drill steel)
7/0	. . . . .	. . . . .	0.500	. . . . .	0.490 0	0.004		
6/0	0.580 0	. . . . .	0.468 75	. . . . .	0.461 5	0.005		
5/0	0.516 5	. . . . .	0.437 5	. . . . .	0.430 5	0.006		
4/0	0.460 0	0.454	0.406 25	. . . . .	0.393 8	0.007		
3/0	0.409 6	0.425	0.375	. . . . .	0.362 5	0.008		
2/0	0.364 8	0.380	0.343 75	. . . . .	0.331 0	0.009		
0	0.324 9	0.340	0.312 5	. . . . .	0.306 5	0.010		
1	0.289 3	0.300	0.281 25	. . . . .	0.283 0	0.011	0.227	0.228 0
2	0.257 6	0.284	0.265 625	. . . . .	0.262 5	0.012	0.219	0.221 0
3	0.229 4	0.259	0.25	0.239 1	0.243 7	0.013	0.212	0.213 0
4	0.204 3	0.238	0.234 375	0.224 2	0.225 3	0.014	0.207	0.209 0
5	0.181 9	0.220	0.218 75	0.209 2	0.207 0	0.016	0.204	0.205 5
6	0.162 0	0.203	0.203 125	0.194 3	0.192 0	0.018	0.201	0.204 0
7	0.144 3	0.180	0.187 5	0.179 3	0.177 0	0.020	0.199	0.201 0
8	0.128 5	0.165	0.171 875	0.164 4	0.162 0	0.022	0.197	6.199 0
9	0.114 5	0.148	0.156 25	0.149 5	0.148 3	0.024	0.194	0.196 0
10	0.101 9	0.134	0.140 625	0.134 5	0.135 0	0.026	0.191	0.193 5
11	0.090 74	0.120	0.125	0.119 6	0.120 5	0.029	0.188	0.191 0
12	0.080 81	0.109	0.109 357	0.104 6	0.105 5		0.185	0.189 0

13	0.071 96	0.095	0.093 75	0.089 7	0.091 5	0.031	0.182	0.185 0
14	0.064 08	0.083	0.078 125	0.074 7	0.080 0	0.033	0.180	0.182 0
15	0.057 07	0.072	0.070 312 5	0.067 3	0.072 0	0.035	0.178	0.180 0
16	0.050 82	0.065	0.062 5	0.059 8	0.062 5	0.037	0.175	0.177 0
17	0.045 26	0.058	0.056 25	0.053 8	0.054 0	0.039	0.172	0.173 0
18	0.040 30	0.049	0.05	0.047 8	0.047 5	0.041	0.168	0.169 5
19	0.035 89	0.042	0.043 75	0.041 8	0.041 0	0.043	0.164	0.166 0
20	0.031 96	0.035	0.037 5	0.035 9	0.034 8	0.045	0.161	0.161 0
21	0.028 46	0.032	0.034 375	0.032 9	0.031 7	0.047	0.157	0.159 0
22	0.025 35	0.028	0.031 25	0.029 9	0.028 6	0.049	0.155	0.157 0
23	0.022 57	0.025	0.028 125	0.026 9	0.025 8	0.051	0.153	0.154 0
24	0.020 10	0.022	0.025	0.023 9	0.023 0	0.055	0.151	0.152 0
25	0.017 90	0.020	0.021 875	0.020 9	0.020 4	0.059	0.148	0.149 5
26	0.015 94	0.018	0.018 75	0.017 9	0.018 1	0.063	0.146	0.147 0
27	0.014 20	0.016	0.017 187 5	0.016 4	0.017 3	0.067	0.143	0.144 0
28	0.012 64	0.014	0.015 625	0.014 9	0.016 2	0.071	0.139	0.140 5
29	0.011 26	0.013	0.014 062 5	0.013 5	0.015 0	0.075	0.134	0.136 0
30	0.010 03	0.012	0.012 5	0.012 0	0.014 0	0.080	0.127	0.128 5
31	0.008 928	0.010	0.010 937 5	0.010 5	0.013 2	0.085	0.120	0.120 0
32	0.007 950	0.009	0.010 156 25	0.009 7	0.012 8	0.090	0.115	0.116 0
33	0.007 080	0.008	0.009 375	0.009 0	0.011 8	0.095	0.112	0.113 0
34	0.006 305	0.007	0.008 593 75	0.008 2	0.010 4	.....	0.110	0.111 0
35	0.005 615	0.005	0.007 812 5	0.007 5	0.009 5	.....	0.108	0.110 0
36	0.005 000	0.004	0.007 031 25	0.006 7	0.009 0	.....	0.106	0.106 5
37	0.004 453	.....	0.006 640 625	0.006 4	0.008 5	.....	0.103	0.104 0
38	0.003 965	.....	0.006 25	0.006 0	0.008 0	.....	0.101	0.101 5
39	0.003 531	.....	.....	.....	0.007 5	.....	0.099	0.099 5
40	0.003 145	.....	.....	.....	0.007 0	.....	0.097	0.098 0

SOURCE: Reynolds Metals Co., Richmond, Virginia.

TABLE 1-18 Properties of Square and Rectangular Structural Steel Tubing†

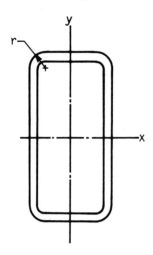

Size, in	Weight, lb/ft	Area A, in²	Radius‡ r, in	I_x, in⁴	I_y, in⁴
$2 \times 2 \times \frac{3}{16}$	4.32	1.27	$\frac{3}{8}$	0.668	
$\frac{1}{4}$	5.41	1.59	$\frac{1}{2}$	0.766	
$3 \times 2 \times \frac{3}{16}$	5.59	1.64	$\frac{3}{8}$	1.24	0.977
$\frac{1}{4}$	7.11	2.09	$\frac{1}{2}$	2.21	1.15
$3 \times 3 \times \frac{3}{16}$	6.87	2.02	$\frac{3}{8}$	2.60	
$\frac{1}{4}$	8.81	2.59	$\frac{1}{2}$	3.16	
$\frac{5}{16}$	10.58	3.11	$\frac{5}{8}$	3.58	
$4 \times 2 \times \frac{3}{16}$	6.87	2.02	$\frac{3}{8}$	3.87	1.29
$\frac{1}{4}$	8.81	2.59	$\frac{1}{2}$	4.69	1.54
$\frac{5}{16}$	10.58	3.11	$\frac{5}{8}$	5.32	1.71
$4 \times 3 \times \frac{3}{16}$	8.15	2.39	$\frac{3}{8}$	5.23	3.34
$\frac{1}{4}$	10.51	3.09	$\frac{1}{2}$	6.45	4.10
$\frac{5}{16}$	12.70	3.73	$\frac{5}{8}$	7.45	4.71
$4 \times 4 \times \frac{3}{16}$	9.42	2.77	$\frac{3}{8}$	6.59	
$\frac{1}{4}$	12.21	3.59	$\frac{1}{2}$	8.22	
$\frac{5}{16}$	14.83	4.36	$\frac{5}{8}$	9.58	
$\frac{3}{8}$	17.27	5.08	$\frac{3}{4}$	10.7	
$\frac{1}{2}$	21.63	6.36	1	12.3	
$5 \times 3 \times \frac{1}{4}$	12.21	3.59	$\frac{1}{2}$	11.3	5.05
$\frac{5}{16}$	14.83	4.36	$\frac{5}{8}$	13.2	5.85
$\frac{3}{8}$	17.27	5.08	$\frac{3}{4}$	14.7	6.48
$\frac{1}{2}$	21.63	6.36	1	16.9	7.33
$5 \times 4 \times \frac{1}{4}$	13.91	4.09	$\frac{1}{2}$	14.1	9.98
$\frac{5}{16}$	16.96	4.98	$\frac{5}{8}$	16.6	11.7
$\frac{3}{8}$	19.82	5.83	$\frac{3}{4}$	18.7	13.2

TABLE 1-18 Properties of Square and Rectangular Structural Steel Tubing† (*Continued*)

Size, in	Weight, lb/ft	Area A, in^2	Radius‡ r, in	I_x, in^4	I_y, in^4
$5 \times 5 \times \frac{1}{4}$	15.62	4.59	$\frac{1}{2}$	16.9	
$\frac{5}{16}$	19.08	5.61	$\frac{5}{8}$	20.1	
$\frac{3}{8}$	22.37	6.58	$\frac{3}{4}$	22.8	
$\frac{1}{2}$	28.43	8.36	1	27.0	
$6 \times 3 \times \frac{1}{4}$	13.91	4.09	$\frac{1}{2}$	17.9	6.00
$\frac{5}{16}$	16.96	4.98	$\frac{5}{8}$	21.1	6.98
$\frac{3}{8}$	19.82	5.83	$\frac{3}{4}$	23.8	7.78
$6 \times 4 \times \frac{1}{4}$	15.62	4.59	$\frac{1}{2}$	22.1	11.7
$\frac{5}{16}$	19.08	5.61	$\frac{5}{8}$	26.2	13.8
$\frac{3}{8}$	22.37	6.58	$\frac{3}{4}$	29.7	15.6
$\frac{1}{2}$	28.43	8.36	1	35.3	18.4
$6 \times 6 \times \frac{1}{4}$	19.02	5.59	$\frac{1}{2}$	30.3	
$\frac{5}{16}$	23.34	6.86	$\frac{5}{8}$	36.3	
$\frac{3}{8}$	27.48	8.08	$\frac{3}{4}$	41.6	
$\frac{1}{2}$	35.24	10.4	1	50.5	
$8 \times 4 \times \frac{5}{16}$	23.34	6.86	$\frac{5}{8}$	53.9	18.1
$\frac{3}{8}$	27.48	8.08	$\frac{3}{4}$	61.9	20.6
$\frac{1}{2}$	35.24	10.4	1	75.1	24.6
$8 \times 6 \times \frac{5}{16}$	27.59	8.11	$\frac{5}{8}$	72.4	46.4
$\frac{3}{8}$	32.58	9.58	$\frac{3}{4}$	83.7	53.5
$\frac{1}{2}$	42.05	12.4	1	103.	65.7
$8 \times 8 \times \frac{5}{16}$	31.84	9.36	$\frac{5}{8}$	90.9	
$\frac{3}{8}$	37.69	11.1	$\frac{3}{4}$	106.	
$\frac{1}{2}$	48.85	14.4	1	131.	
$\frac{5}{8}$	59.32	17.4	$1\frac{1}{4}$	153.	

†Size expressed by outside dimensions and wall thickness; other sizes are available (see Ref. [1-2]).
‡Tolerance is three times the wall thickness.

takes all the variation. However, tubing can be ordered using an inside-diameter specification.

1-5 WIRE AND SHEET METAL

Gauge sizes of wire and sheet metal of both ferrous and nonferrous materials are tabulated in Table 1-17. The use of SI units is simpler for such products because it is easier to express thicknesses directly in millimeters.

(*continued on p. 43*)

TABLE 1-19 Properties of American Standard Channels†

Designation	Area A, in^2	t_w, in	b, in	t_f, in	D, in	$\bar{x}$, in	e, in	I_x, in^4	I_y, in^4
C 3 × 4.1	1.26	0.170	1.410	0.273	$\cdots$	0.436	0.461	1.66	0.197
3 × 5	1.47	0.258	1.498	0.273	$\cdots$	0.438	0.392	1.85	0.247
3 × 6	1.76	0.356	1.596	0.273	$\cdots$	0.455	0.322	2.07	0.305
C 4 × 5.4	1.59	0.184	1.584	0.296	$\cdots$	0.457	0.502	3.85	0.319
4 × 7.25	2.13	0.321	1.721	0.296	$\frac{5}{8}$	0.459	0.386	4.59	0.433
C 5 × 6.7	1.97	0.190	1.750	0.320	$\cdots$	0.484	0.552	7.49	0.479
5 × 9	2.64	0.325	1.885	0.320	$\frac{5}{8}$	0.478	0.427	8.90	0.632

					D				
C	6 × 8.2	2.40	0.200	1.920	0.343	$\frac{5}{16}$	0.511	0.599	13.1	0.693
	6 × 10.5	3.09	0.314	2.034	0.343	$\frac{5}{16}$	0.499	0.486	15.2	0.866
	6 × 13	3.83	0.437	2.157	0.343	$\frac{5}{16}$	0.514	0.380	17.4	1.05
C	7 × 9.8	2.87	0.210	2.090	0.366	$\frac{5}{16}$	0.540	0.647	21.3	0.968
	7 × 12.25	3.60	0.314	2.194	0.366	$\frac{5}{16}$	0.525	0.538	24.2	1.17
	7 × 14.75	4.33	0.419	2.299	0.366	$\frac{5}{16}$	0.532	0.441	27.2	1.38
C	8 × 11.5	3.38	0.220	2.260	0.390	$\frac{3}{4}$	0.571	0.697	32.6	1.32
	8 × 13.75	4.04	0.303	2.343	0.390	$\frac{3}{4}$	0.553	0.604	36.1	1.53
	8 × 18.75	5.51	0.487	2.527	0.390	$\frac{3}{4}$	0.565	0.431	44.0	1.98
C	9 × 13.4	3.94	0.233	2.433	0.413	$\frac{3}{4}$	0.601	0.743	47.9	1.76
	9 × 15	4.41	0.285	2.485	0.413	$\frac{3}{4}$	0.586	0.682	51.0	1.93
	9 × 20	5.88	0.448	2.648	0.413	$\frac{3}{4}$	0.583	0.515	60.9	2.42
C	10 × 15.3	4.49	0.240	2.600	0.436	$\frac{3}{4}$	0.634	0.796	67.4	2.28
	10 × 20	5.88	0.379	2.739	0.436	$\frac{3}{4}$	0.606	0.637	78.9	2.81
	10 × 25	7.35	0.526	2.886	0.436	$\frac{3}{4}$	0.617	0.494	91.2	3.36
	10 × 30	8.82	0.673	3.033	0.436	$\frac{3}{4}$	0.649	0.369	103	3.94
C	12 × 20.7	6.09	0.282	2.942	0.501	$\frac{7}{8}$	0.698	0.870	129	3.88
	12 × 25	7.35	0.387	3.047	0.501	$\frac{7}{8}$	0.674	0.746	144	4.47
	12 × 30	8.82	0.510	3.170	0.501	$\frac{7}{8}$	0.674	0.618	162	5.14
C	15 × 33.9	9.96	0.400	3.400	0.650	1	0.787	0.896	315	8.13
	15 × 40	11.8	0.520	3.520	0.650	1	0.777	0.767	349	9.23
	15 × 50	14.7	0.716	3.716	0.650	1	0.798	0.583	404	11.0

†The designation is the channel depth and the unit weight in pounds per foot; D = diameter of maximum flange fastener, and e = location of shear center.

SOURCE: Ref. [1-1]. All the sizes listed here are generally available in aluminum alloys. For these, the unit weight is obtained by multiplying the area by 0.829.

TABLE 1-20 Properties of Angles†

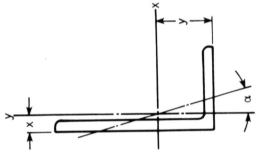

Size, in	w, lb/ft	Area A, in²	y, in	I_x, in⁴	x, in	I_y, in⁴	Tan α
L 2 × 2 × $\frac{1}{8}$	1.65	0.484	0.546	0.190	0.546	0.190	1.000
× $\frac{3}{16}$	2.44	0.715	0.569	0.272	0.569	0.272	1.000
× $\frac{1}{4}$	3.19	0.938	0.592	0.348	0.592	0.348	1.000
× $\frac{5}{16}$	3.92	1.15	0.614	0.416	0.614	0.416	1.000
× $\frac{3}{8}$	4.7	1.36	0.636	0.479	0.636	0.479	1.000
L 2½ × 2 × $\frac{3}{16}$	2.75	0.809	0.764	0.509	0.514	0.291	0.631
× $\frac{1}{4}$	3.62	1.06	0.787	0.654	0.537	0.372	0.626
× $\frac{5}{16}$	4.5	1.31	0.809	0.788	0.559	0.446	0.620
× $\frac{3}{8}$	5.3	1.55	0.831	0.912	0.581	0.514	0.614
L 2½ × 2½ × $\frac{3}{16}$	3.07	0.902	0.694	0.547	. . .	. . .	1.000
× $\frac{1}{4}$	4.10	1.19	0.717	0.703	. . .	. . .	1.000
× $\frac{5}{16}$	5.00	1.46	0.740	0.849	. . .	. . .	1.000
× $\frac{3}{8}$	5.9	1.73	0.762	0.984	. . .	. . .	1.000

Size							
L 3 × 2 × 3/16	3.07	0.902	0.970	0.842	0.470	0.307	0.446
L 3 × 2 × 1/4	4.1	1.19	0.993	1.09	0.493	0.392	0.440
× 5/16	5.0	1.46	1.02	1.32	0.516	0.470	0.435
× 3/8	5.9	1.73	1.04	1.53	0.539	0.543	0.428
L 3 × 2½ × 3/16	3.39	0.996	0.888	0.907	0.638	0.577	0.688
× 1/4	4.5	1.31	0.911	1.17	0.661	0.743	0.684
× 3/8	6.6	1.92	0.956	1.66	0.706	1.04	0.676
L 3 × 3 × 3/16	3.71	1.09	0.820	0.962	....	....	1.000
× 1/4	4.9	1.44	0.842	1.24	....	....	1.000
× 5/16	6.1	1.78	0.865	1.51	....	....	1.000
× 3/8	7.2	2.11	0.888	1.76	....	....	1.000
× 1/2	9.4	2.75	0.932	2.22	....	....	1.000
L 3½ × 2½ × 1/4	4.9	1.44	1.11	1.80	0.614	0.777	0.506
× 5/16	6.1	1.78	1.14	2.19	0.637	0.939	0.501
× 3/8	7.2	2.11	1.16	2.56	0.660	1.09	0.496
L 3½ × 3 × 1/4	5.4	1.56	1.04	1.91	0.785	1.30	0.727
× 5/16	6.6	1.93	1.06	2.33	0.808	1.58	0.724
L 3½ × 3 × 3/8	7.9	2.30	1.08	2.72	0.830	1.85	0.721
L 3½ × 3½ × 1/4	5.8	1.69	0.968	2.01	....	....	1.000
× 5/16	7.2	2.09	0.990	2.45	....	....	1.000
× 3/8	8.5	2.48	1.01	2.87	....	....	1.000
L 4 × 3 × 1/4	5.8	1.69	1.24	2.77	0.736	1.36	0.558
× 5/16	7.2	2.09	1.26	3.38	0.759	1.65	0.554
× 3/8	8.5	2.48	1.28	3.96	0.782	1.92	0.551
× 1/2	11.1	3.25	1.33	5.05	0.827	2.42	0.543
L 4 × 3½ × 1/4	6.2	1.81	1.16	2.91	0.909	2.09	0.759
× 5/16	7.7	2.25	1.18	3.56	0.932	2.55	0.757
× 3/8	9.1	2.67	1.21	4.18	0.955	2.95	0.755
× 1/2	11.9	3.50	1.25	5.32	1.00	3.79	0.750

TABLE 1-20 Properties of Angles† (*Continued*)

Size, in	w, lb/ft	Area A, in²	y, in	I_x, in⁴	x, in	I_y, in⁴	Tan α
L 4 × 4 × 1/4	6.6	1.94	1.09	3.04	.....	.....	1.000
× 5/16	8.2	2.40	1.12	3.71	.....	.....	1.000
× 3/8	9.8	2.86	1.14	4.36	.....	.....	1.000
× 1/2	12.8	3.75	1.18	5.56	.....	.....	1.000
L 4 × 4 × 5/8	15.7	4.61	1.23	6.66	.....	.....	1.000
× 3/4	18.6	5.44	1.27	7.67	.....	.....	1.000
L 5 × 3 × 1/4	6.6	1.94	1.66	5.11	0.657	1.44	0.371
× 5/16	8.2	2.40	1.68	6.26	0.681	1.75	0.368
× 3/8	9.8	2.86	1.70	7.37	0.704	2.04	0.364
× 1/2	12.8	3.75	1.75	9.45	0.750	2.58	0.357
L 5 × 3½ × 5/16	8.7	2.56	1.59	6.60	0.838	2.72	0.489
× 3/8	10.4	3.05	1.61	7.78	0.861	3.18	0.486
× 1/2	13.6	4.00	1.66	9.99	0.906	4.05	0.479
× 3/4	19.8	5.81	1.75	13.9	0.996	5.55	0.464
L 5 × 5 × 5/16	10.3	3.03	1.37	7.42	.....	.....	1.000
× 3/8	12.3	3.61	1.39	8.74	.....	.....	1.000
× 1/2	16.2	4.75	1.43	11.3	.....	.....	1.000
× 3/4	23.6	6.94	1.52	15.7	.....	.....	1.000
× 7/8	27.2	7.98	1.57	17.8	.....	.....	1.000
L 6 × 3½ × 5/16	9.8	2.87	2.01	10.9	0.763	2.85	0.352
× 3/8	11.7	3.42	2.04	12.9	0.787	3.34	0.350

Size†						
L 6 × 4 × 3/8	12.3	3.61	1.94	13.5	0.941	4.90	0.446
× 1/2	16.2	4.75	1.99	17.4	0.987	6.27	0.440
× 5/8	20.0	5.86	2.03	21.1	1.03	7.52	0.435
× 3/4	23.6	6.94	2.08	24.5	1.08	8.68	0.428
L 6 × 6 × 3/8	14.9	4.36	1.64	15.4	....	....	1.000
× 1/2	19.6	5.75	1.68	19.9	....	....	1.000
× 5/8	24.2	7.11	1.73	24.2	....	....	1.000
× 3/4	28.7	8.44	1.78	28.2	....	....	1.000
× 7/8	33.1	9.73	1.82	31.9	....	....	1.000
× 1	37.4	11.0	1.86	35.5	....	....	1.000
L 7 × 4 × 3/8	13.6	3.98	2.37	20.6	0.870	5.10	0.340
× 1/2	17.9	5.25	2.42	26.7	0.917	6.53	0.335
× 3/4	26.2	7.69	2.51	37.8	1.01	9.05	0.324
L 8 × 4 × 1/2	19.6	5.75	2.86	38.5	0.859	6.74	0.267
× 3/4	28.7	8.44	2.95	54.9	0.953	9.36	0.258
× 1	37.4	11.0	3.05	69.6	1.05	11.6	0.247
L 8 × 6 × 1/2	23.0	6.75	2.47	44.3	1.47	21.7	0.558
× 3/4	33.8	9.94	2.56	63.4	1.56	30.7	0.551
× 1	44.2	13.0	2.65	80.8	1.65	38.8	0.543
L 8 × 8 × 1/2	26.4	7.75	2.19	48.6	....	....	1.000
× 5/8	32.7	9.61	2.23	59.4	....	....	1.000
× 3/4	38.9	11.4	2.28	69.7	....	....	1.000
× 7/8	45.0	13.2	2.32	79.6	....	....	1.000
× 1	51.0	15.0	2.37	89.0	....	....	1.000
× 1 1/8	56.9	16.7	2.41	98.0	....	....	1.000

†Size is the length of each leg and the thickness; unit weight for steel is w.
SOURCE: Ref. [1-1]. Angles up to 6 in inclusive are also available in aluminum alloys. For these, the unit weight is obtained by multiplying the area by 0.829. Sizes in structural steel larger than those listed are available on special order.

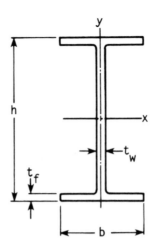

TABLE 1-21 Properties of W Shapes†

Designation	Area A, in²	h, in	t_w, in	b, in	t_f, in	I_x, in⁴	I_y, in⁴
W 4 × 13	3.83	4.16	0.280	4.060	0.345	11.3	3.86
W 5 × 16	4.68	5.01	0.240	5.000	0.360	21.3	7.51
5 × 19	5.54	5.15	0.270	5.030	0.430	26.2	9.13
W 6 × 9	2.68	5.90	0.170	3.940	0.215	16.4	2.19
6 × 12	3.55	6.03	0.230	4.000	0.280	22.1	2.99
6 × 16	4.74	6.28	0.260	4.030	0.405	32.1	4.43
6 × 15	4.43	5.99	0.230	5.990	0.260	29.1	9.32
6 × 20	5.87	6.20	0.260	6.020	0.365	41.4	13.3
6 × 25	7.34	6.38	0.320	6.080	0.455	53.4	17.1
W 8 × 10	2.96	7.89	0.170	3.940	0.205	30.8	2.09
8 × 13	3.84	7.99	0.230	4.000	0.255	39.6	2.73
8 × 15	4.44	8.11	0.245	4.015	0.315	48.0	3.41
8 × 18	5.26	8.14	0.230	5.25	0.330	61.9	7.97
8 × 21	6.16	8.28	0.250	5.27	0.400	75.3	9.77
8 × 24	7.08	7.93	0.245	6.495	0.400	82.8	18.3
8 × 28	8.25	8.06	0.285	6.535	0.465	98.0	21.7
8 × 31	9.13	8.00	0.285	7.995	0.435	110	37.1
8 × 35	10.3	8.12	0.310	8.020	0.495	127	42.6
8 × 40	11.7	8.25	0.360	8.070	0.560	146	49.1
8 × 48	14.1	8.50	0.400	8.110	0.685	184	60.9
8 × 58	17.1	8.75	0.510	8.220	0.810	228	75.1
8 × 67	19.7	9.00	0.570	8.280	0.935	272	88.6
W 10 × 12	3.54	9.87	0.190	3.960	0.210	53.8	2.18
10 × 15	4.41	9.99	0.230	4.000	0.270	68.9	2.89
10 × 17	4.99	10.11	0.240	4.010	0.330	81.9	3.56
10 × 19	5.62	10.24	0.250	4.020	0.395	96.3	4.29
W 10 × 22	6.49	10.17	0.240	5.75	0.360	118	11.4
10 × 26	7.61	10.33	0.260	5.770	0.440	144	14.1
10 × 30	8.84	10.47	0.300	5.810	0.510	170	16.7
W 10 × 33	9.71	9.73	0.290	7.960	0.435	170	36.6
10 × 39	11.5	9.92	0.315	7.985	0.530	209	45.0
10 × 45	13.3	10.10	0.350	8.020	0.620	248	53.4

†The designation is the nominal depth, and the unit weight for steel is in pounds per foot. Larger sizes are available from W 10 × 49 to W 36 × 300. See Ref. [1-2]. Some of the sizes 8 in and under are available in aluminum alloys which are then called H sections.

TABLE 1-22 Properties of S Shapes†

Designation	Area A, in²	h, in	t_w, in	b, in	t_f, in	D, in	I_x, in⁴	I_y, in⁴
S 3 × 5.7	1.67	3.00	0.170	2.330	0.260	...	2.52	0.455
S 3 × 7.5	2.21	3.00	0.349	2.509	0.260	...	2.93	0.586
S 4 × 7.7	2.26	4.00	0.193	2.663	0.293	...	6.08	0.764
S 4 × 9.5	2.79	4.00	0.326	2.796	0.293	...	6.79	0.903
S 5 × 10	2.94	5.00	0.214	3.004	0.326	...	12.3	1.22
S 5 × 14.75	4.34	5.00	0.494	3.284	0.326	...	15.2	1.67
S 6 × 12.5	3.67	6.00	0.232	3.332	0.359	...	22.1	1.82
S 6 × 17.25	5.07	6.00	0.465	3.565	0.359	$\frac{5}{8}$	26.3	2.31
S 7 × 15.3	4.50	7.00	0.252	3.662	0.392	$\frac{5}{8}$	36.7	2.64
S 7 × 20	5.88	7.00	0.450	3.860	0.392	$\frac{5}{8}$	42.4	3.17

Notes • Drawings • Ideas

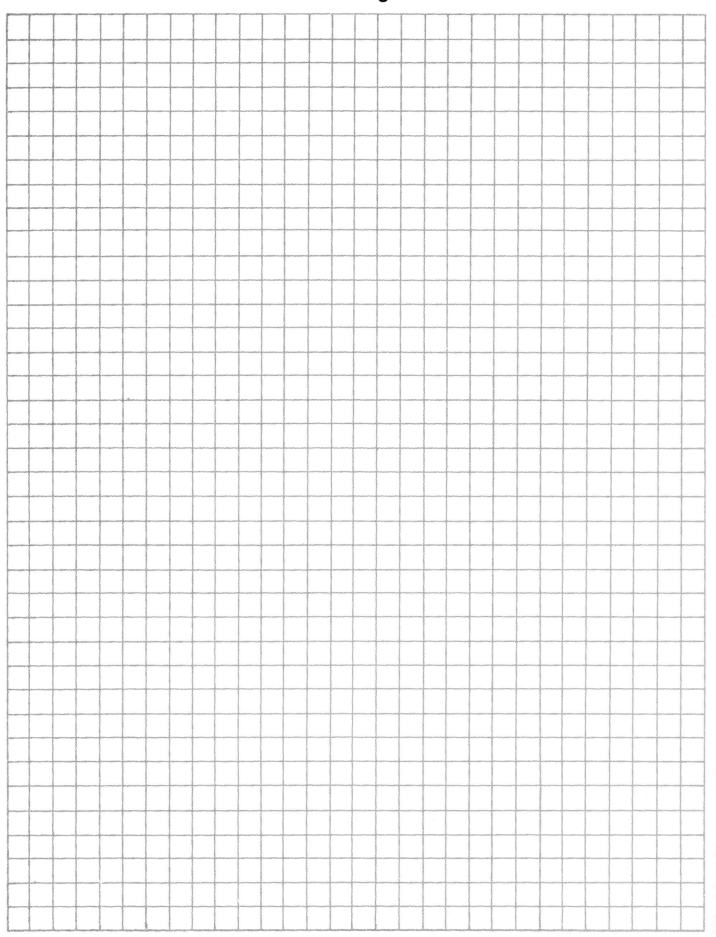

TABLE 1-22 Properties of S Shapes† (*Continued*)

Designation	Area A, in^2	h, in	t_w, in	b, in	t_f, in	D, in	I_x, in^4	I_y, in^4
S 8 × 18.4	5.41	8.00	0.271	4.001	0.426	3/4	57.6	3.73
8 × 23	6.77	8.00	0.441	4.171	0.426	3/4	64.9	4.31
S 10 × 25.4	7.46	10.00	0.311	4.661	0.491	3/4	124	6.79
10 × 35	10.3	10.00	0.594	4.944	0.491	3/4	147	8.36
S 12 × 31.8	9.35	12.00	0.350	5.000	0.544	3/4	218	9.36
12 × 35	10.3	12.00	0.428	5.078	0.544	3/4	229	9.87
S 12 × 40.8	12.0	12.00	0.462	5.252	0.659	3/4	272	13.6
12 × 50	14.7	12.00	0.687	5.477	0.659	3/4	305	15.7
S 15 × 42.9	12.6	15.00	0.411	5.501	0.622	3/4	447	14.4
15 × 50	14.7	15.00	0.550	5.640	0.622	3/4	486	15.7
S 18 × 54.7	16.1	18.00	0.461	6.001	0.691	7/8	804	20.8
18 × 70	20.6	18.00	0.711	6.251	0.691	7/8	926	24.1
S 20 × 66	19.4	20.00	0.505	6.255	0.795	7/8	1190	27.7
20 × 75	22.0	20.00	0.635	6.385	0.795	7/8	1280	29.8
S 20 × 86	25.3	20.30	0.660	7.060	0.920	1	1580	46.8
20 × 96	28.2	20.30	0.800	7.200	0.920	1	1670	50.2
S 24 × 80	23.5	24.00	0.500	7.000	0.870	1	2100	42.2
24 × 90	26.5	24.00	0.625	7.125	0.870	1	2250	44.9
24 × 100	29.3	24.00	0.745	7.245	0.870	1	2390	47.4

†The designation is the nominal depth and the unit weight for steel is in pounds per foot; D = diameter of maximum flange fastener.

SOURCE: Ref. [1–2]. Many of the sizes in this table up to and including 12 in are also available in aluminum alloys. Multiply the area by 0.829 to get the weight of these shapes.

Notes ▪ Drawings ▪ Ideas

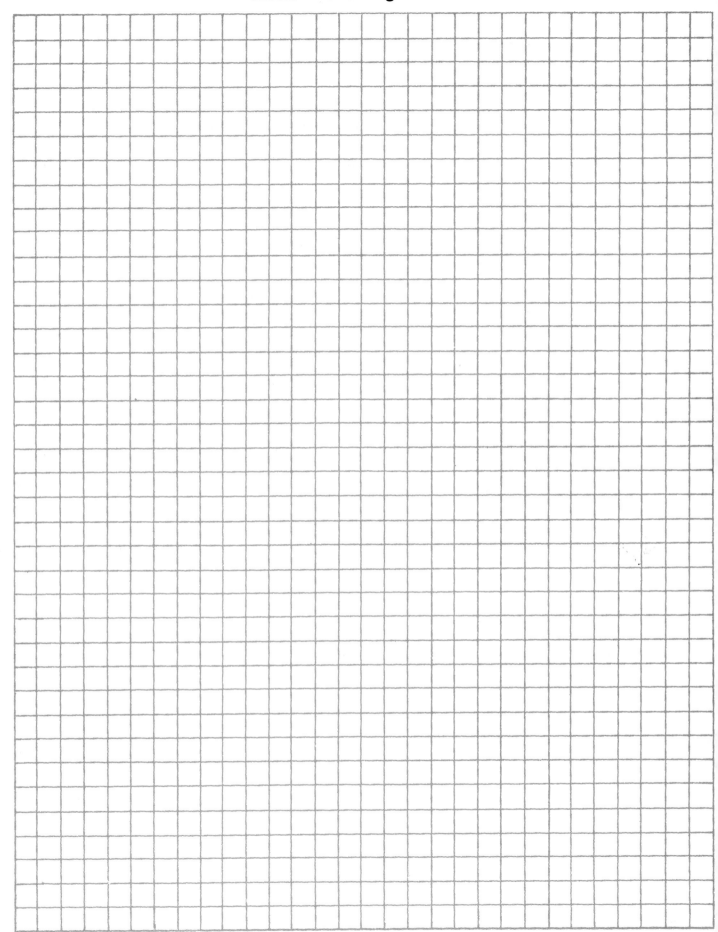

1-6 STRUCTURAL SHAPES

An assortment of various shapes used in structural steel works and their sizes and properties are tabulated in Tables 1-18 to 1-22. These are probably the most useful sizes for machine-design purposes, but other sizes are available or can be obtained on special order. Generally, aluminum shapes are available in a larger range of sizes, especially the smaller ones.

REFERENCES

1-1 *Steel Products Manual*, American Iron and Steel Institute, Washington, D.C.

1-2 *Manual of Steel Construction*, American Institute of Steel Construction, Inc., Chicago, Illinois.

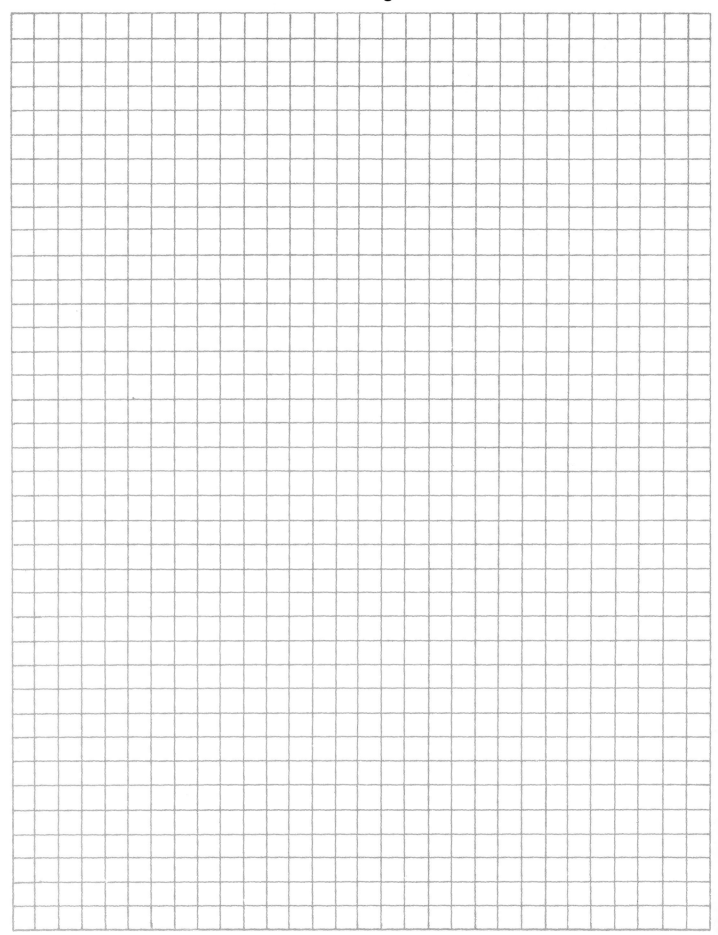

chapter **2**
STRESS

JOSEPH E. SHIGLEY
Professor Emeritus
University of Michigan
Ann Arbor, Michigan

2-1 DEFINITIONS AND NOTATION

The general two-dimensional stress element in Fig. 2-1a shows two normal stresses σ_x and σ_y, both positive, and two shear stresses τ_{xy} and τ_{yx}, positive also. The element is in static equilibrium, and hence $\tau_{xy} = \tau_{yx}$. The stress state depicted by the figure is called *plane* or *biaxial stress*.

Figure 2-1b shows an element face whose normal makes and angle ϕ to the x axis. It can be shown that the stress components σ and τ acting on this face are given by the equations

$$\sigma = \frac{\sigma_x + \sigma_y}{2} + \frac{\sigma_x - \sigma_y}{2} \cos 2\phi + \tau_{xy} \sin 2\phi \qquad (2\text{-}1)$$

$$\tau = -\frac{\sigma_x - \sigma_y}{2} \sin 2\phi + \tau_{xy} \cos 2\phi \qquad (2\text{-}2)$$

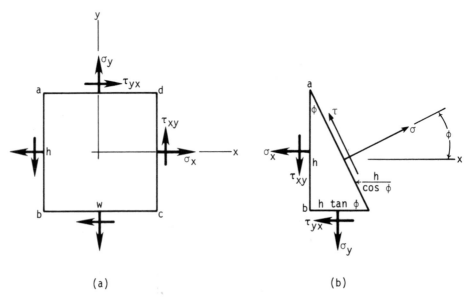

(a)   (b)

FIG. 2-1 Notation for two-dimensional stress. *(From Applied Mechanics of Materials, by Joseph E. Shigley. Copyright © 1976 by McGraw-Hill, Inc. Used with permission of the McGraw-Hill Book Company.)*

It can be shown that when the angle ϕ is varied in Eq. (2-1) the normal stress σ has two extreme values. These are called the *principal stresses,* and they are given by the equation

$$\sigma_1, \sigma_2 = \frac{\sigma_x + \sigma_y}{2} \pm \left[\frac{\sigma_x - \sigma_y}{2} + \tau_{xy}^2 \right]^{1/2} \tag{2-3}$$

The corresponding values of ϕ are called the *principal directions.* These directions can be obtained from

$$2\phi = \tan^{-1} \frac{2\tau_{xy}}{\sigma_x - \sigma_y} \tag{2-4}$$

The shear stresses are always zero when the element is aligned in the principal directions.

It also turns out that the shear stress τ in Eq. (2-2) has two extreme values. These and the angles at which they occur may be found from

$$\tau_1, \tau_2 = \pm \left[\left(\frac{\sigma_x - \sigma_y}{2} \right)^2 + \tau_{xy}^2 \right]^{1/2} \tag{2-5}$$

$$2\phi = \tan^{-1} - \frac{\sigma_x - \sigma_y}{2\tau_{xy}} \tag{2-6}$$

The two normal stresses are equal when the element is aligned in the directions given by Eq. (2-6).

The act of referring stress components to another reference system is called *transformation of stress.* Such transformations are easier to visualize, and to solve, using a *Mohr's circle diagram.* In Fig. 2-2 we create a $\sigma\tau$ coordinate system with normal stresses plotted as the ordinates. On the abscissa, tensile (positive) normal stresses are plotted to the right of the origin O, and compression (negative) normal stresses are plotted to the left. The sign convention for shear stresses is that clockwise (cw) shear stresses are plotted *above* the abscissa and counterclockwise (ccw) shear stresses are plotted *below.*

The stress state of Fig. 2-1a is shown on the diagram in Fig. 2-2. Points A and C represent σ_x and σ_y, respectively, and point E is midway between them. Distance AB is τ_{xy} and distance CD is τ_{yx}. The circle of radius ED is *Mohr's circle.* This circle passes through the principal stresses at F and G and through the extremes of the shear stresses at H and I. It is important to observe that an extreme of the shear stress may *not* be the same as the maximum.

2-1-1 Programming

To program a Mohr's circle solution, plan on using a rectangular-to-polar conversion subroutine. Now notice, in Fig. 2-2, that $(\sigma_x - \sigma_y)/2$ is the base of a right triangle, τ_{xy} is the ordinate, and the hypotenuse is an extreme of the shear stress. Thus the conversion routine can be used to output both the angle 2ϕ and the extreme value of the shear stress.

As shown in Fig. 2-2, the principal stresses are found by adding and subtracting the extreme value of the shear stress to the term $(\sigma_x + \sigma_y)/2$. It is wise to ensure, in your programming, that the angle ϕ indicates the angle *from* the x axis *to* the direction of the stress component of interest; generally, the angle ϕ is considered positive when measured in the ccw direction.

Notes · Drawings · Ideas

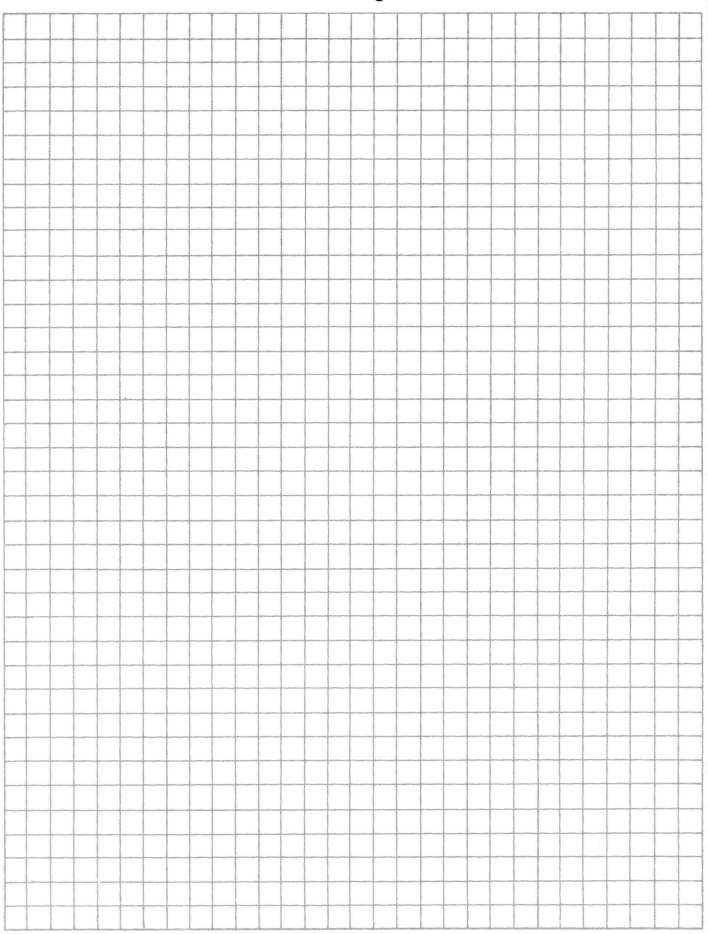

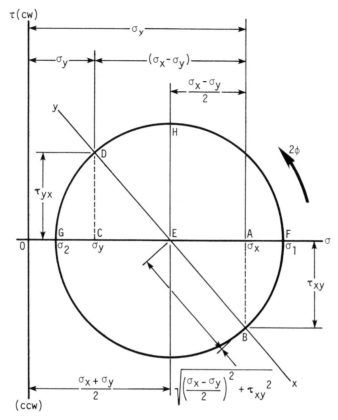

FIG. 2-2 Mohr's circle diagram for plane stress. (*From Applied Mechanics of Materials, by Joseph E. Shigley. Copyright © 1976 by McGraw-Hill, Inc. Used with permission of the McGraw-Hill Book Company.*)

2-2 TRIAXIAL STRESS

The general three-dimensional stress element in Fig. 2-3*a* has three normal stresses σ_x, σ_y, and σ_z, all shown as positive, and six shear-stress components, also shown positive. The element is in static equilibrium, and hence

$$\tau_{xy} = \tau_{yx} \qquad \tau_{yz} = \tau_{zy} \qquad \tau_{zx} = \tau_{xz}$$

Note that the first subscript is the coordinate normal to the element face; the second subscript designates the axis parallel to the shear-stress component. The negative faces of the element will have shear stresses acting in the opposite direction; these are also considered as positive.

As shown in Fig. 2-3*b*, there are three principal stresses for triaxial stress states. These three are obtained from a solution of the equation

$$\sigma^3 - (\sigma_x + \sigma_y + \sigma_z)\sigma^2 + (\sigma_x\sigma_y + \sigma_x\sigma_z + \sigma_y\sigma_z - \tau_{xy}^2 - \tau_{yz}^2 - \tau_{zx}^2)\sigma$$

$$- (\sigma_x\sigma_y\sigma_z + 2\tau_{xy}\tau_{yz}\tau_{zx} - \sigma_x\tau_{yz}^2 - \sigma_y\tau_{zx}^2 - \sigma_z\tau_{xy}^2) = 0 \qquad (2\text{-}7)$$

In plotting Mohr's circles for triaxial stress, arrange the principal stresses in the order $\sigma_1 > \sigma_2 > \sigma_3$, as in Fig. 2-3*b*. It can be shown that the stress coordinates $\sigma\tau$ for any arbitrarily located plane will always lie on or *inside* the largest circle or on

Notes ▪ Drawings ▪ Ideas

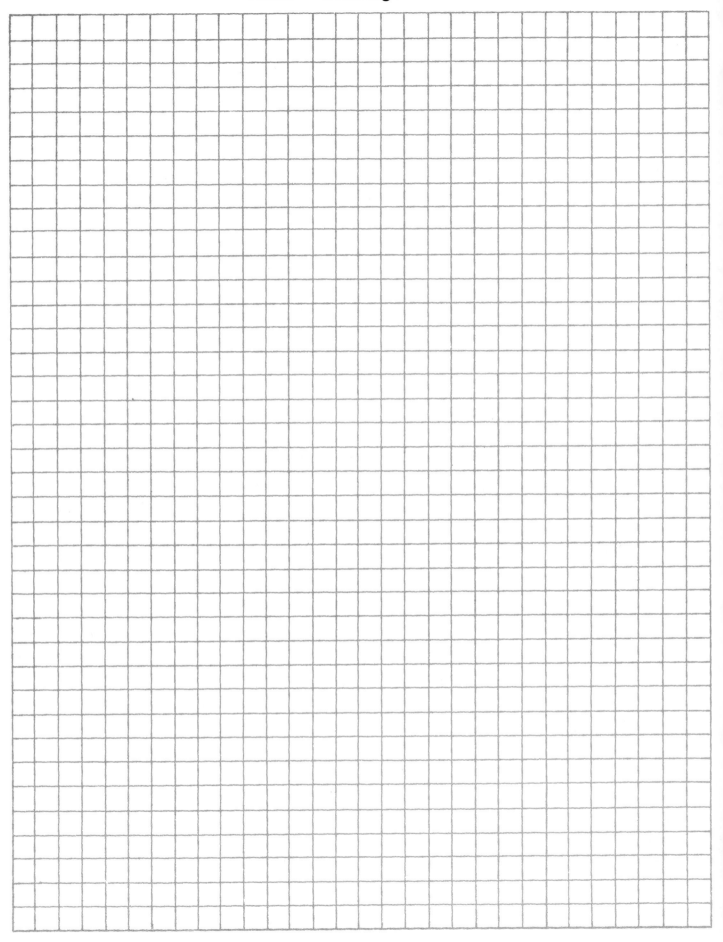

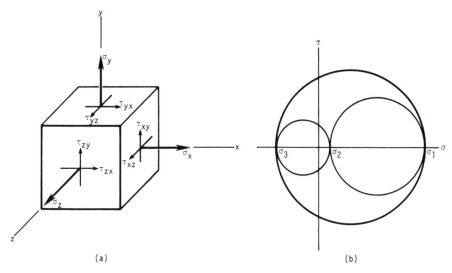

FIG. 2-3 (*a*) General triaxial stress element; (*b*) Mohr's circles for triaxial stress.

or *outside* the two smaller circles. The figure shows that the maximum shear stress is always

$$\tau_{max} = \frac{\sigma_1 - \sigma_3}{2} \tag{2-8}$$

when the normal stresses are arranged so that $\sigma_1 > \sigma_2 > \sigma_3$.

2-3 STRESS-STRAIN RELATIONS

The stresses due to loading described as *pure tension, pure compression,* and *pure shear* are

$$\sigma = \frac{F}{A} \qquad \tau = \frac{F}{A} \tag{2-9}$$

where F is positive for tension and negative for compression and the word *pure* means that there are no other complicating effects. In each case the stress is assumed to be uniform, which requires that:

- The member is straight and of a homogeneous material.
- The line of action of the force is through the centroid of the section.
- There is no discontinuity or change in cross section near the stress element.
- In the case of compression, there is no possibility of buckling.

Unit engineering strain ϵ, often called simply *unit strain,* is the elongation or deformation of a member subjected to pure axial loading per unit of original length. Thus

$$\epsilon = \frac{\delta}{l_0} \tag{2-10}$$

where δ = total strain
  l_0 = unstressed or original length

Shear strain γ is the change in a right angle of a stress element due to pure shear.

Hooke's law states that, within certain limits, the stress in a material is proportional to the strain which produced it. Materials which regain their original shape and dimensions when a load is removed are called *elastic materials*. Hooke's law is expressed in equation form as

$$\sigma = E\epsilon \qquad \tau = G\gamma \qquad (2\text{-}11)$$

where E = the *modulus of elasticity,* and G = the *modulus of ridigity,* also called the *shear modulus of elasticity.*

Poisson demonstrated that, within the range of Hooke's law, a member subjected to uniaxial loading exhibits both an axial strain and a lateral strain. These are related to each other by the equation

$$\nu = -\frac{\text{lateral strain}}{\text{axial strain}} \qquad (2\text{-}12)$$

where ν is called *Poisson's ratio.*

The three constants given by Eqs. (2-11) and (2-12) are often called *elastic constants.* They have the relationship

$$E = 2G(1 + \nu) \qquad (2\text{-}13)$$

By combining Eqs. (2-9), (2-10), and (2-11) it is easy to show that

$$\delta = \frac{Fl}{AE} \qquad (2\text{-}14)$$

which gives the total deformation of a member subjected to axial tension or compression.

A solid round bar subjected to a *pure* twisting moment or torsion has a shear stress that is zero at the center and maximum at the surface. The appropriate equations are

$$\tau = \frac{T\rho}{J} \qquad \tau_{\max} = \frac{Tr}{J} \qquad (2\text{-}15)$$

where T = torque
  ρ = radius to stress element
  r = radius of bar
  J = second moment of area (polar)

The total angle of twist of such a bar, in radians, is

$$\theta = \frac{Tl}{GJ} \qquad (2\text{-}16)$$

where l = length of the bar. For the shear stress and angle of twist of other cross sections see Table 2-1.

2-3-1 Principal Unit Strains

For a bar in uniaxial tension or compression, the principal strains are

$$\epsilon_1 = \frac{\sigma_1}{E} \qquad \epsilon_2 = -\nu\epsilon_1 \qquad \epsilon_3 = -\nu\epsilon_1 \qquad (2\text{-}17)$$

Notice that the stress state is uniaxial, but the strains are triaxial.

For triaxial stress the principal strains are

$$\epsilon_1 = \frac{\sigma_1}{E} - \frac{\nu\sigma_2}{E} - \frac{\nu\sigma_3}{E}$$

$$\epsilon_2 = \frac{\sigma_2}{E} - \frac{\nu\sigma_1}{E} - \frac{\nu\sigma_3}{E} \qquad (2\text{-}18)$$

$$\epsilon_3 = \frac{\sigma_3}{E} - \frac{\nu\sigma_1}{E} - \frac{\nu\sigma_2}{E}$$

These equations can be solved for the principal stresses; the results are

$$\sigma_1 = \frac{E\epsilon_1(1-\nu) + \nu E(\epsilon_2 + \epsilon_3)}{1 - \nu - 2\nu^2}$$

$$\sigma_2 = \frac{E\epsilon_2(1-\nu) + \nu E(\epsilon_1 + \epsilon_3)}{1 - \nu - 2\nu^2} \qquad (2\text{-}19)$$

$$\sigma_3 = \frac{E\epsilon_3(1-\nu) + \nu E(\epsilon_1 + \epsilon_2)}{1 - \nu - 2\nu^2}$$

The biaxial stress-strain relations can easily be obtained from Eqs. (2-18) and (2-19) by equating one of the principal stresses to zero.

2-3-2 Plastic Strain

It is important to observe that all the preceding relations are valid only when the material obeys Hooke's law.

Some materials, when stressed in the plastic region, exhibit a behavior quite similar to that given by Eq. (2-11). For these materials, the appropriate equation is

$$\bar{\sigma} = K\varepsilon^n \qquad (2\text{-}20)$$

where $\bar{\sigma}$ = true stress
K = strength coefficient
ε = true plastic strain
n = strain-hardening exponent

The relations for the true stress and true strain are

$$\bar{\sigma} = \frac{F_i}{A_i} \qquad \varepsilon = \ln\frac{l_i}{l_0} \qquad (2\text{-}21)$$

where A_i and l_i are, respectively, the instantaneous values of the area and length of a bar subjected to a load F_i. Note that the areas in Eq. (2-9) are the original or

TABLE 2-1 Torsional Stress and Angular Deflection of Various Sections†

Sectional shape	Shape constant	Shear stress
1. Solid round	$K = \dfrac{\pi d^4}{32}$	$\tau_{\max} = \dfrac{16T}{\pi d^3}$
2. Round tube	$K = \dfrac{\pi(d_o^4 - d_i^4)}{32}$	$\tau_{\max} = \dfrac{16Td_o}{\pi(d_o^4 - d_i^4)}$
3. Square [10-1]	$K = \dfrac{h^4}{7.2}$	$\tau_{\max} = \dfrac{4.8T}{h^3}$

4. Square tube, generous fillets [10-2]

$$K = t(h - t)^3$$

$$\tau \cong \frac{T}{2t(h - t)^2}$$

5. Rectangle [10-1]

$$K = \frac{bh^3}{A}$$

$$A = 3 + 1.462\frac{h}{b} + 2.976\left(\frac{h}{b}\right)^2 - 0.238\left(\frac{h}{b}\right)^3$$

$$h \leq b$$

$$\tau_{max} = \frac{T(3b + 1.8h)}{b^2h^2}$$

6. Rectangular tube, generous fillets [10-2]

$$K = \frac{2t(b - t)^2(h - t)^2}{b + h - 2t}$$

$$\tau \cong \frac{T}{2t(b - t)(h - t)}$$

TABLE 2-1 Torsional Stress and Angular Deflection of Various Sections† (*Continued*)

Sectional shape	Shape constant	Shear stress
7. Hexagon [10-1]	$K = \dfrac{h^4}{8.8}$	$\tau_{\max} = \dfrac{5.7T}{h^3}$

†Deflection is $\theta = Tl/KG$ in rad, where T = torque, l = length, K = shape constant, and G = modulus of rigidity.

56

unstressed areas; the subscript zero was omitted, as is customary. The relations between true and engineering (nominal) stresses and strains are

$$\bar{\sigma} = \sigma \exp \epsilon \qquad \varepsilon = \ln(\epsilon + 1) \qquad (2\text{-}22)$$

2-4 FLEXURE

Figure 2-4a shows a member loaded in flexure by a number of forces F and supported by reactions R_1 and R_2 at the ends. At point C a distance x from R_1 we can write

$$\Sigma M_C = \Sigma M_{\text{ext}} + M = 0 \qquad (2\text{-}23)$$

where $\Sigma M_{\text{ext}} = -xR_1 + c_1F_1 + c_2F_2$ and is called the *external moment* at section C. The term M, called the *internal* or *resisting moment,* is shown in its positive direc-

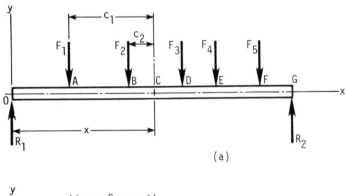

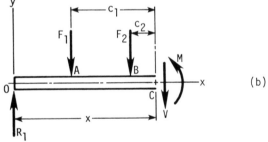

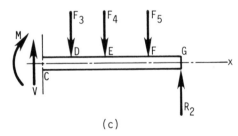

FIG. 2-4 Shear and moment. *(From Applied Mechanics of Materials, by Joseph E. Shigley. Copyright © 1976 by McGraw-Hill, Inc. Used with permission of the McGraw-Hill Book Company.)*

Notes ▪ Drawings ▪ Ideas

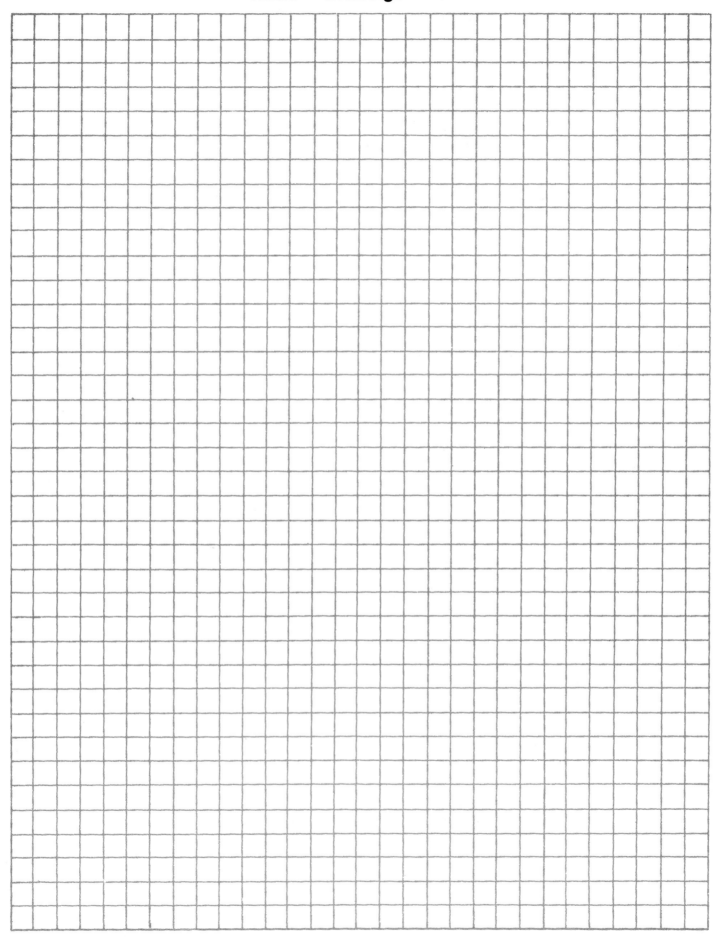

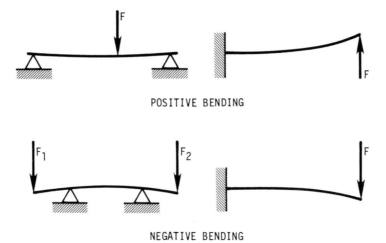

POSITIVE BENDING

NEGATIVE BENDING

FIG. 2-5 Sign conventions for bending. *(From Applied Mechanics of Materials, by Joseph E. Shigley. Copyright © 1976 by McGraw-Hill, Inc. Used with permission of the McGraw-Hill Book Company.)*

tion in both parts *b* and *c* of Fig. 2-4. Figure 2-5 shows that a positive moment causes the top surface of a beam to be concave. A negative moment causes the top surface to be convex with one or both ends curved downward.

A similar relation can be defined for shear at section *C:*

$$\Sigma F_y = \Sigma F_{ext} + V = 0 \qquad (2\text{-}24)$$

where $\Sigma F_{ext} = R_1 - F_1 - F_2$ and is called the *external shear force* at C. The term *V*, called the *internal shear force,* is shown in its positive direction in both parts *b* and *c* of Fig. 2-4.

Figure 2-6 illustrates an application of these relations to obtain a set of shear and moment diagrams.

The previous relations can be expressed in a more general form as

$$V = \frac{dM}{dx} \qquad (2\text{-}25)$$

If the flexure is caused by a distributed load,

$$\frac{dV}{dx} = \frac{d^2M}{dx^2} = -w \qquad (2\text{-}26)$$

where *w* = a downward-acting load in units of force per unit length. A more general load distribution can be expressed as

$$q = \lim_{\Delta x \to 0} \frac{\Delta F}{\Delta x}$$

where *q* is called the *load intensity;* thus $q = -w$ in Eq. (2-26). Two useful facts can be learned by integrating Eqs. (2-25) and (2-26). The first is

$$\int_{V_A}^{V_B} dV = \int_{x_A}^{x_B} q \, dx = V_B - V_A \qquad (2\text{-}27)$$

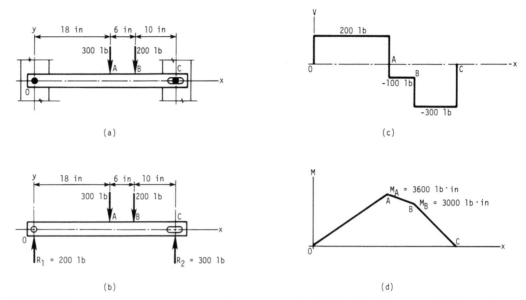

FIG. 2-6 (*a*) View showing how ends are secured; (*b*) loading diagram; (*c*) shear-force diagram; (*d*) bending-moment diagram. *(From Applied Mechanics of Materials, by Joseph E. Shigley. Copyright © 1976 by McGraw-Hill, Inc. Used with permission of the McGraw-Hill Book Company.)*

which states that *the area under the loading function between x_A and x_B is the same as the change in the shear force from A to B.* Also,

$$\int_{M_A}^{M_B} dM = \int_{x_A}^{x_B} V \, dx = M_B - M_A \tag{2-28}$$

which states that *the area of the shear-force diagram between x_A and x_B is the same as the change in moment from A to B.*

Figure 2-7 distinguishes between the *neutral axis of a section* and the *neutral axis of a beam,* both of which are often referred to simply as the *neutral axis.* The assumptions used in deriving flexural relations are

- The material is isotropic and homogeneous.
- The member is straight.
- The material obeys Hooke's law.
- The cross section is constant along the length of the member.
- There is an axis of symmetry in the plane of bending (see Fig. 2-7).
- During pure bending (zero shear force), plane cross sections remain plane.

The *flexural formula* is

$$\sigma_x = -\frac{My}{I} \tag{2-29}$$

for the section of Fig. 2-7. The formula states that a normal compression stress σ_x occurs on a fiber at distance y from the neutral axis when a *positive moment M* is applied. The moment of inertia I is really the *second moment of area;* formulas may be found in Chap. 1.

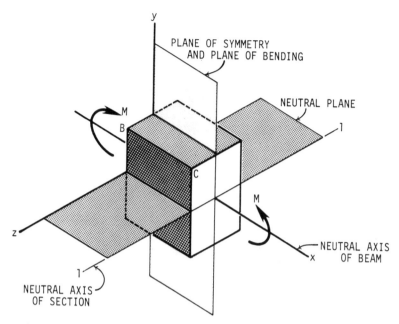

FIG. 2-7 The meaning of the term *neutral axis*. Note the difference between the *neutral axis of the section* and the *neutral axis of the beam*. *(From Applied Mechanics of Materials, by Joseph E. Shigley. Copyright © 1976 by McGraw-Hill, Inc. Used with permission of the McGraw-Hill Book Company.)*

The maximum flexural stress occurs at $y_{\max} = c$ at the outer surface of the beam. This stress is often written in the three forms

$$\sigma = \frac{Mc}{I} \qquad \sigma = \frac{M}{I/c} \qquad \sigma = \frac{M}{Z} \qquad (2\text{-}30)$$

where Z is called the *section modulus*. Equation (2-30) can also be used for beams having unsymmetrical sections provided that the plane of bending coincides with one of the two principal axes of the section.

When shear forces are present, as in Fig. 2-6c, a member in flexure will also experience shear stresses as given by the equation

$$\tau = \frac{VQ}{Ib} \qquad (2\text{-}31)$$

where b = section width, and Q = first moment of a vertical face about the neutral axis and is

$$Q = \int_{y_1}^{c} y\, dA \qquad (2\text{-}32)$$

For a rectangular section

$$Q = \int_{y_1}^{c} y\, dA = b \int_{y_1}^{c} y\, dy = \frac{b}{2}(c^2 - y_1^2)$$

Notes · Drawings · Ideas

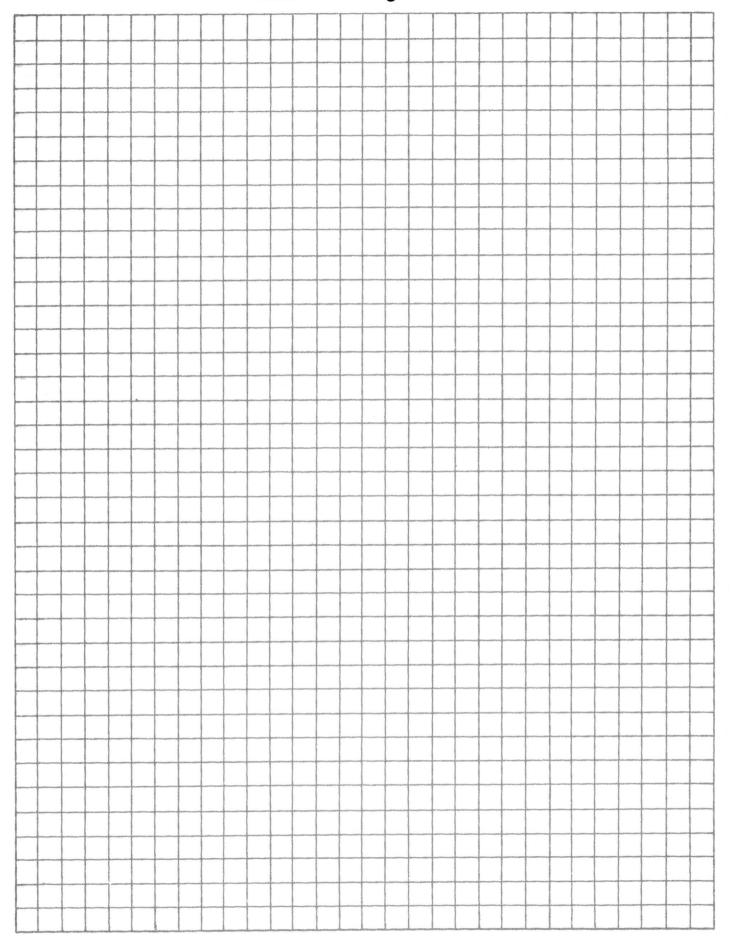

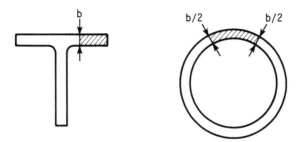

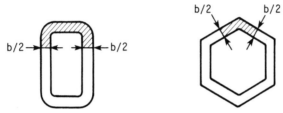

FIG. 2-8 Correct way to measure dimension b to determine shear stress for various sections. *(From Applied Mechanics of Materials, by Joseph E. Shigley. Copyright © 1976 by McGraw-Hill, Inc. Used with permission of the McGraw-Hill Book Company.)*

Substituting this value of Q into Eq. (2-31) gives

$$\tau = \frac{V}{2I}(c^2 - y_1^2)$$

Using $I = Ac^2/3$, we learn that

$$\tau = \frac{3V}{2A}\left(1 - \frac{y_1^2}{c^2}\right) \tag{2-33}$$

The value of b for other sections is measured as shown in Fig. 2-8.

In determining shear stress in a beam, the dimension b is not always measured parallel to the neutral axis. The beam sections shown in Fig. 2-8 show how to measure b in order to compute the static moment Q. It is the tendency of the shaded area to slide relative to the unshaded area, which causes the shear stress.

Shear flow q is defined by the equation

$$q = \frac{VQ}{I} \tag{2-34}$$

where q is in force units per unit length of the beam at the section under consideration. So shear flow is simply the shear force per unit length at the section defined by $y = y_1$. When the shear flow is known, the shear stress is determined by the equation

$$\tau = \frac{q}{b} \tag{2-35}$$

2-5   STRESSES DUE TO TEMPERATURE

A *thermal stress* is caused by the existence of a *temperature gradient* in a member. A *temperature stress* is created in a member when it is *constrained* so as to prevent expansion or contraction due to temperature change.

2-5-1   Temperature Stresses

These stresses are found by assuming the member is not constrained and then computing the stresses required to cause it to assume its original dimensions. If the temperature of an unrestrained member is uniformly increased, the member expands and the normal strain is

$$\epsilon_x = \epsilon_y = \epsilon_z = \alpha(\Delta T) \tag{2-36}$$

where ΔT = temperature change, and α = *coefficient of linear expansion*. The coefficient of linear expansion increases to some extent with temperature. Some mean values for various materials are shown in Table 2-2.

Figure 2-9 illustrates two examples of temperature stresses. For the bar in Fig. 2-9*a*.

$$\sigma_x = -\alpha(\Delta T)E \qquad \sigma_y = \sigma_z = -\nu\sigma_x \tag{2-37}$$

The stresses in the flat plate of Fig. 2-9*b* are

$$\sigma_x = \sigma_y = -\frac{\alpha(\Delta T)E}{1 - \nu} \qquad \sigma_z = -\nu\sigma_x \tag{2-38}$$

TABLE 2-2 Coefficients of Linear Expansion

Material	Celsius scale		Fahrenheit scale	
	$10^6\alpha$	°C	$10^6\alpha$	°F
Aluminum	24.0	20–100	13.4	68–212
Aluminum	26.7	20–300	14.9	68–572
Brass (cast)	18.75	0–100	10.4	32–212
Brass (wire)	19.3	0–100	10.7	32–212
Brass (spring)	19.8	25–300	11.0	77–572
Cast iron	10.6	40	5.9	104
Carbon steel	10.8	40	6.0	104
Carbon steel	11.5	100–200	6.4	212–392
Carbon steel	15	300–400	8.3	572–752
Magnesium (cast)	27.0	20–100	15.0	68–212
Nickel steel (10%)	13.0	20	7.2	68
Stainless steel (hardened)	9.6	20–100	5.3	68–212
Stainless steel (hardened)	9.8	20–200	5.5	68–392
Stainless steel (annealed)	10.3	20–100	5.7	68–212
Stainless steel (annealed)	10.7	20–200	6.0	68–392

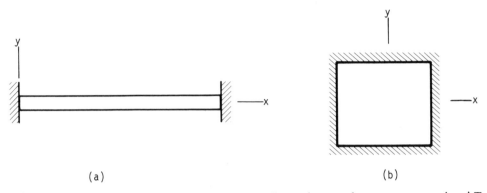

(a)　　　　　　　　　　(b)

FIG. 2-9 Examples of temperature stresses. In each case the temperature rise ΔT is uniform throughout. (a) Straight bar with ends restrained; (b) flat plate with edges restrained.

2-5-2 Thermal Stresses

Heating of the top surface of the restrained member in Fig. 2-10a causes end moment of

$$M = \frac{\alpha(\Delta T)EI}{h} \qquad (2\text{-}39)$$

and maximum bending stresses of

$$\sigma_x = \pm \frac{\alpha(\Delta T)E}{2} \qquad (2\text{-}40)$$

with compression of the top surface. If the constraints are removed, the bar will curve to a radius

$$r = \frac{h}{\alpha(\Delta T)}$$

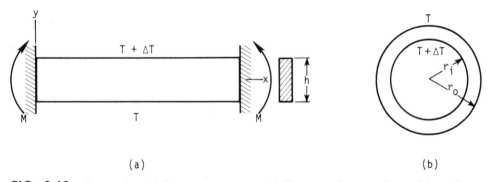

(a)　　　　　　　　　　(b)

FIG. 2-10 Examples of thermal stresses. (a) Rectangular member with ends restrained (temperature difference between top and bottom results in end moments and bending stresses); (b) thick-walled tube has maximum stresses in tangential and longitudinal directions.

Notes ▪ Drawings ▪ Ideas

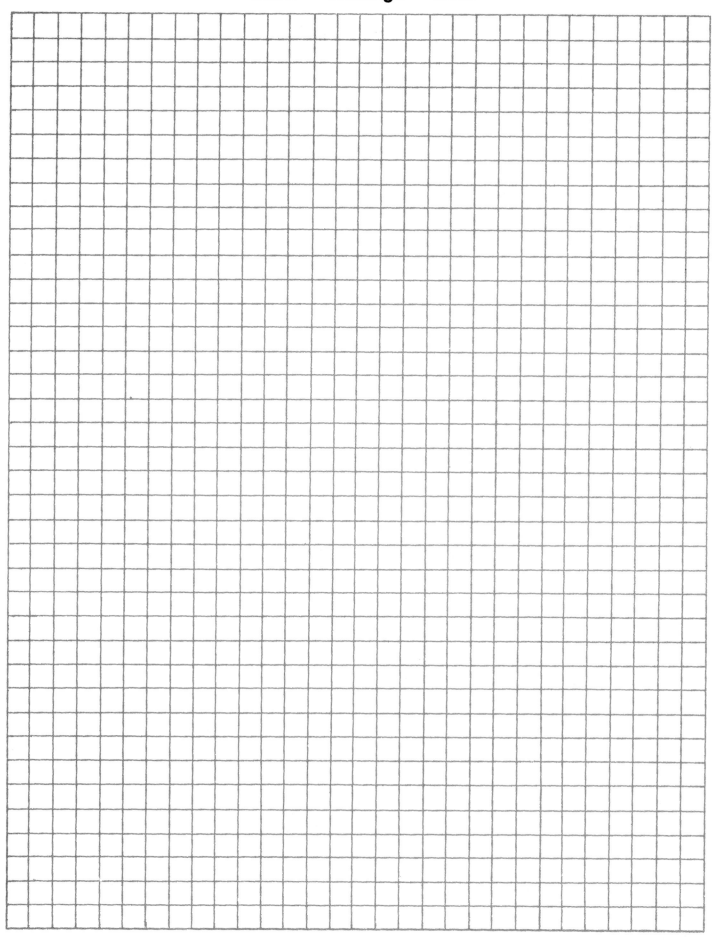

The thick-walled tube of Fig. 2-10b with a hot interior surface has tangential and longitudinal stresses in the outer and inner surfaces of magnitude

$$\sigma_{lo} = \sigma_{to} = \frac{\alpha(\Delta T)E}{2(1 - \nu)\ln(r_o/r_i)}\left[1 - \frac{2r_i^2\ln(r_o/r_i)}{r_o^2 - r_i^2}\right] \qquad (2\text{-}41)$$

$$\sigma_{li} = \sigma_{ti} = \frac{-\alpha(\Delta T)E}{2(1 - \nu)\ln(r_o/r_i)}\left[1 - \frac{2r_o^2\ln(r_o/r_i)}{r_o^2 - r_i^2}\right] \qquad (2\text{-}42)$$

where subscripts i and o refer to the inner and outer radii, respectively, and the subscripts t and l refer to the tangential (circumferential) and longitudinal directions. Radial stresses of lesser magnitude will also exist, although not at the inner or outer surfaces.

If the tubing of Fig. 2-10b is thin, then the inner and outer stresses are equal, although opposite, and are

$$\sigma_{lo} = \sigma_{to} = \frac{\alpha(\Delta T)E}{2(1 - \nu)}$$

$$\sigma_{li} = \sigma_{ti} = -\frac{\alpha(\Delta T)E}{2(1 - \nu)} \qquad (2\text{-}43)$$

at points not too close to the tube ends.

2-6 CONTACT STRESSES

When two elastic bodies having curved surfaces are pressed against each other, the initial point or line of contact changes into area contact, because of the deformation, and a three-dimensional state of stress is induced in both bodies. The shape of the contact area was originally deduced by Hertz, who assumed that the curvature of the two bodies could be approximated by second-degree surfaces. For such bodies, the contact area was found to be an ellipse. Reference [2-3] contains a comprehensive bibliography.

As indicated in Fig. 2-11, there are four special cases in which the contact area is a circle. For these four cases the maximum pressure occurs at the center of the contact area and is

$$p_o = \frac{3F}{2\pi a^2} \qquad (2\text{-}44)$$

where a = the radius of the contact area, and F = the normal force pressing the two bodies together.

In Fig. 2-11 the x and y axes are in the plane of the contact area and the z axis is normal to this plane. The maximum stresses occur on this axis, they are principal stresses, and their values for all four cases in Fig. 2-11 are

$$\sigma_x = \sigma_y = -p_o\left\{\left(1 - \frac{z}{a}\tan^{-1}\frac{1}{z/a}\right)(1 + \nu) - \frac{1}{2(1 + z^2/a^2)}\right\} \qquad (2\text{-}45)$$

$$\sigma_z = \frac{-p_o}{1 + z^2/a^2} \qquad (2\text{-}46)$$

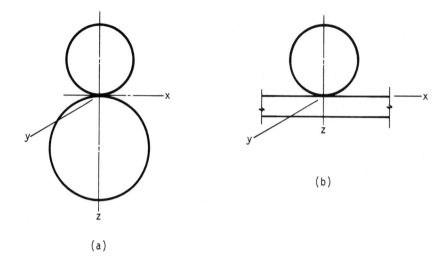

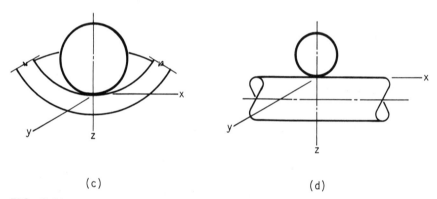

FIG. 2-11   Contacting bodies having a circular area. (*a*) Two spheres; (*b*) sphere and plate; (*c*) sphere and spherical socket; (*d*) crossed cylinders of equal diameters.

These equations are plotted in Fig. 2-12 together with the two shear stresses τ_{xz} and τ_{yz}. Note that $\tau_{xy} = 0$ because $\sigma_x = \sigma_y$.

The radii a of the contact circles depend on the geometry of the contacting bodies. For two spheres, each having the same diameter d, or for two crossed cylinders, each having the diameter d, and in each case with like materials, the radius is

$$a = \left(\frac{3Fd}{8} \frac{1 - \nu^2}{E} \right)^{1/3} \tag{2-47}$$

where ν and E are the elastic constants.

For two spheres of unlike materials having diameters d_1 and d_2, the radius is

$$a = \left[\frac{3F}{8} \frac{d_1 d_2}{d_1 + d_2} \left(\frac{1 - \nu_1^2}{E_1} + \frac{1 - \nu_2^2}{E_2} \right) \right]^{1/3} \tag{2-48}$$

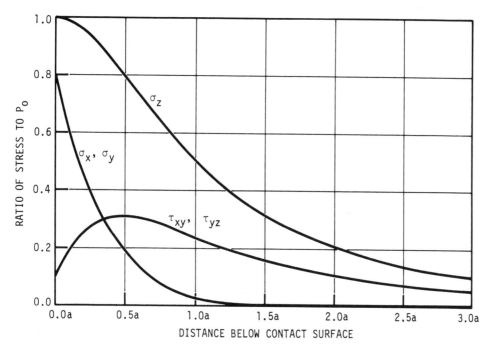

FIG. 2-12 Magnitude of the stress components on the z axis below the surface as a function of the maximum pressure. Note that the two shear-stress components are maximum slightly below the surface. The chart is based on a Poisson's ratio of 0.30.

For a sphere of diameter d and a flat plate of unlike materials, the radius is

$$a = \left[\frac{3Fd}{8} \left(\frac{1 - \nu_1^2}{E_1} + \frac{1 - \nu_2^2}{E_2} \right) \right]^{1/3} \tag{2-49}$$

For a sphere of diameter d_1 and a spherical socket of diameter d_2 of unlike materials, the radius is

$$a = \left[\frac{3F}{8} \frac{d_1 d_2}{d_2 - d_1} \left(\frac{1 - \nu_1^2}{E_1} + \frac{1 - \nu_2^2}{E_2} \right) \right]^{1/3} \tag{2-50}$$

Contacting cylinders with parallel axes subjected to a normal force have a rectangular contact area. We specify an xy plane coincident with the contact area with the x axis parallel to the cylinder axes. Then using a right-handed coordinate system, the stresses along the z axis are maximum and are

$$\sigma_x = -2\nu p_o \left[\left(1 + \frac{z^2}{b^2} \right)^{1/2} - \frac{z}{b} \right] \tag{2-51}$$

$$\sigma_y = -p_o \left[\left(2 - \frac{1}{1 + z^2/b^2} \right) \left(1 + \frac{z^2}{b^2} \right)^{1/2} - \frac{2z}{b} \right] \tag{2-52}$$

$$\sigma_z = \frac{-p_o}{(1 + z^2/b^2)^{1/2}} \tag{2-53}$$

where the maximum pressure occurs at the origin of the coordinate system in the contact zone and is

$$p_o = \frac{2F}{\pi bl} \qquad (2\text{-}54)$$

where l = the length of the contact zone measured parallel to the cylinder axes, and b = the half width. Equations (2-51) to (2-53) give the principal stresses. These equations are plotted in Fig. 2-13. The corresponding shear stresses can be found from a Mohr's circle; they are plotted in Fig. 2-14. Note that the maximum is either τ_{xz} or τ_{yz} depending on the depth below the contact surface.

The half width b depends on the geometry of the contacting cylinders. The following cases arise most frequently: Two cylinders of equal diameter and of the same material have a half width of

$$b = \left(\frac{2Fd}{\pi l} \frac{1 - \nu^2}{E} \right)^{1/2} \qquad (2\text{-}55)$$

For two cylinders of unequal diameter and unlike materials, the half width is

$$b = \left[\frac{2F}{\pi l} \frac{d_1 d_2}{d_1 + d_2} \left(\frac{1 - \nu_1^2}{E_1} + \frac{1 - \nu_2^2}{E_2} \right) \right]^{1/2} \qquad (2\text{-}56)$$

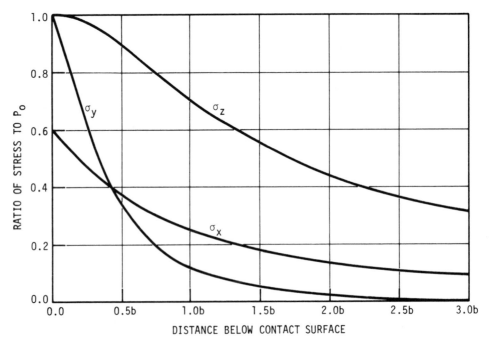

FIG. 2-13 Magnitude of the principal stresses on the z axis below the surface as a function of the maximum pressure for contacting cylinders. Based on a Poisson's ratio of 0.30.

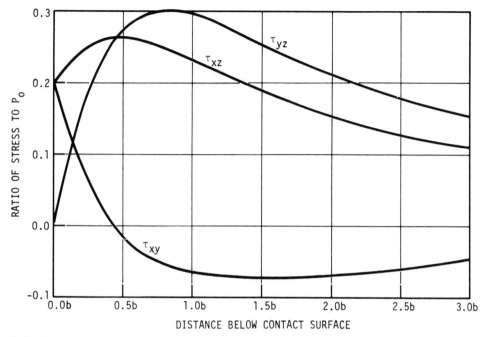

FIG. 2-14 Magnitude of the three shear stresses computed from Fig. 2-13.

For a cylinder of diameter d in contact with a flat plate of unlike material, the result is

$$b = \left[\frac{2Fd}{\pi l} \left(\frac{1 - \nu_1^2}{E_1} + \frac{1 - \nu_2^2}{E_2} \right) \right]^{1/2} \tag{2-57}$$

The half width for a cylinder of diameter d_1 pressing against a cylindrical socket of diameter d_2 of unlike material is

$$b = \left[\frac{2F}{\pi l} \frac{d_1 d_2}{d_2 - d_1} \left(\frac{1 - \nu_1^2}{E_1} + \frac{1 - \nu_2^2}{E_2} \right) \right]^{1/2}$$

REFERENCES

2-1 F. R. Shanley, *Strength of Materials,* McGraw-Hill, Inc., New York, 1957, p. 509.

2-2 R. J. Roark, *Formulas for Stress and Strain,* 4th ed., McGraw-Hill, Inc., New York, 1965, pp. 194–196.

2-3 J. L. Lubkin, "Contact Problems," in W. Flugge (ed.), *Handbook of Engineering Mechanics,* McGraw-Hill, Inc., New York, 1962, pp. 42-10 to 42-12.

Notes ▪ Drawings ▪ Ideas

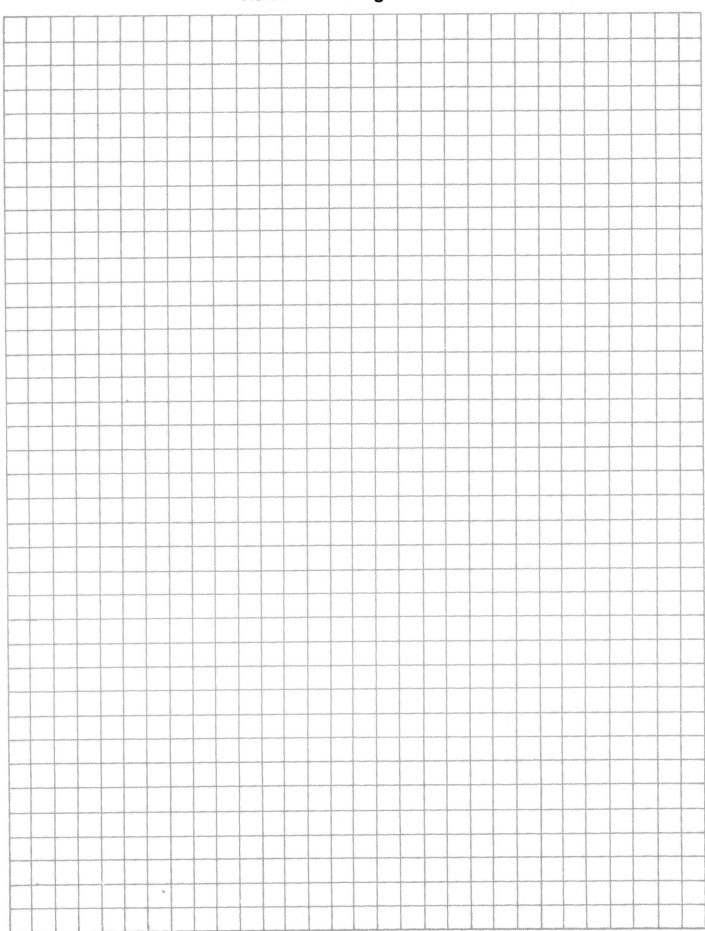

chapter 3
DEFLECTION

JOSEPH E. SHIGLEY

Professor Emeritus
The University of Michigan
Ann Arbor, Michigan

CHARLES R. MISCHKE, Ph.D., P.E.

Professor of Mechanical Engineering
Iowa State University
Ames, Iowa

GLOSSARY OF SYMBOLS

a	Dimension
A	Area
b	Dimension
C	Constant
D, d	Diameter
E	Young's modulus
F	Force
G	Shear modulus
I	Second moment of area
J	Second polar moment of area
k	Spring rate
K	Constant
ℓ	Length
M	Moment
$M(I)$	Moment relation, $(M/EI)_i$
N	Number
q	Unit load
Q	Fictitious force
R	Support reaction
T	Torque
U	Strain energy
V	Shear force
w	Unit weight
W	Total weight
x	Coordinate
y	Coordinate
δ	Deflection
θ	Slope, torsional deflection
ϕ	An integral
ψ	An integral

3-1 STIFFNESS OR SPRING RATE

The *spring rate* (also called *stiffness* or *scale*) of a body or ensemble of bodies is defined as the partial derivative of force (torque) with respect to colinear displacement (rotation). For a helical tension or compression spring,

$$F = \frac{d^4 G y}{8 D^3 N} \quad \text{thus} \quad k = \frac{\partial F}{\partial y} = \frac{d^4 G}{8 D^3 N} \tag{3-1}$$

where D = mean coil diameter
   d = wire diameter
   N = number of active turns

In a round bar subject to torsion,

$$T = \frac{GJ\theta}{\ell} \quad \text{thus} \quad k = \frac{\partial T}{\partial \theta} = \frac{GJ}{\ell} \tag{3-2}$$

and the tensile force in an elongating bar of any cross section is

$$F = \frac{AE\delta}{\ell} \quad \text{thus} \quad k = \frac{\partial F}{\partial \delta} = \frac{AE}{\ell} \tag{3-3}$$

If k is constant, as in these cases, then displacement is said to be linear with respect to force (torque). For contacting bodies with all four radii of curvature finite, the approach of the bodies is proportional to load to the two-thirds power, making the spring rate proportional to load to the one-third power. In hydrodynamic film bearings, the partial derivative would be evaluated numerically by dividing a small change in load by the displacement in the direction of the load.

3-2 DEFLECTION DUE TO BENDING

The relations involved in the bending of beams are well known and are given here for reference purposes as follows:

$$\frac{q}{EI} = \frac{d^4 y}{dx^4} \tag{3-4}$$

$$\frac{V}{EI} = \frac{d^3 y}{dx^3} \tag{3-5}$$

$$\frac{M}{EI} = \frac{d^2 y}{dx^2} \tag{3-6}$$

$$\theta = \frac{dy}{dx} \tag{3-7}$$

$$y = f(x) \tag{3-8}$$

These relations are illustrated by the beam of Fig. 3-1. Note that the x axis is *positive* to the right and the y axis is *positive* upward. All quantities, loading, shear force, support reactions, moment, slope, and deflection have the same sense as y; they are positive if upward, negative if downward.

Notes · Drawings · Ideas

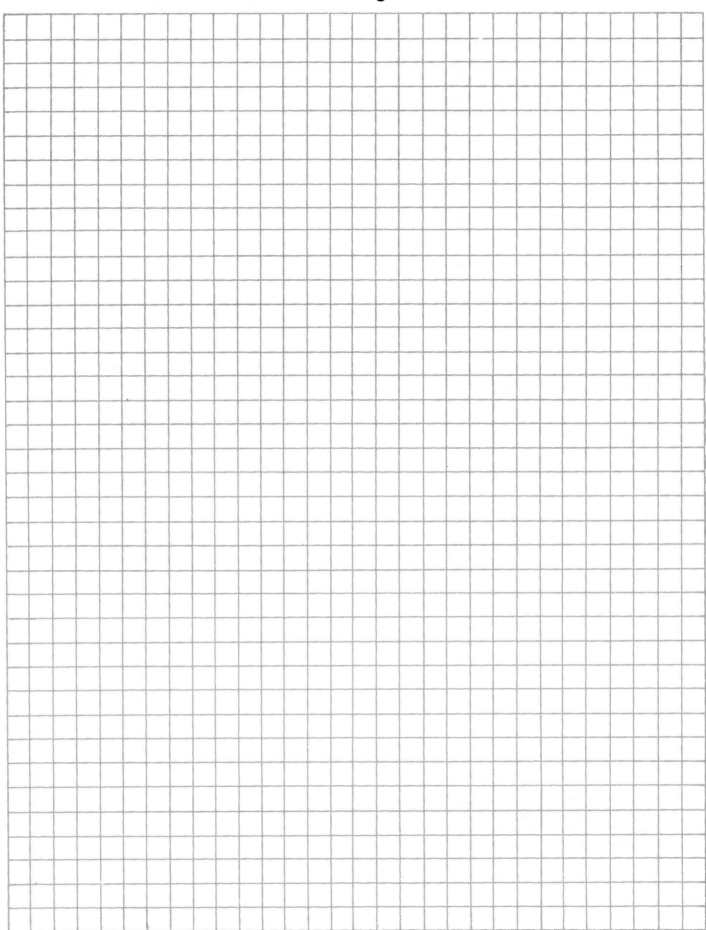

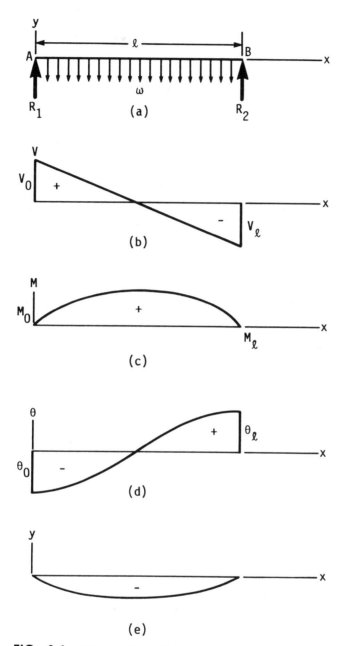

FIG. 3-1 (*a*) Loading diagram showing beam supported at *A* and *B* with uniform load *w* having units of force per unit length, $R_1 = R_2 = w\ell/2$; (*b*) shear-force diagram showing end conditions; (*c*) moment diagram; (*d*) slope diagram; (*e*) deflection diagram.

3-3 PROPERTIES OF BEAMS

Table 3-1 lists a number of useful properties of beams having a variety of loadings. These must all have the same cross section throughout the length, and a linear relation must exist between the force and the deflection. Beams having other loadings can be solved using two or more sets of these relations and the principle of superposition.

TABLE 3-1 Properties of Beams

1. Cantilever—intermediate load

$$R_A = F \qquad M_A = -Fa$$

$$y_B = -\frac{Fa^3}{3EI} \qquad y_C = -\frac{Fa^3}{3EI}\left(1 + \frac{3b}{2a}\right)$$

2. Cantilever—intermediate couple

$$V = 0 \qquad M_A = M$$

$$y_B = -\frac{Ma^2}{2EI} \qquad y_C = -\frac{Ma^2}{2EI}\left(1 + \frac{2b}{a}\right)$$

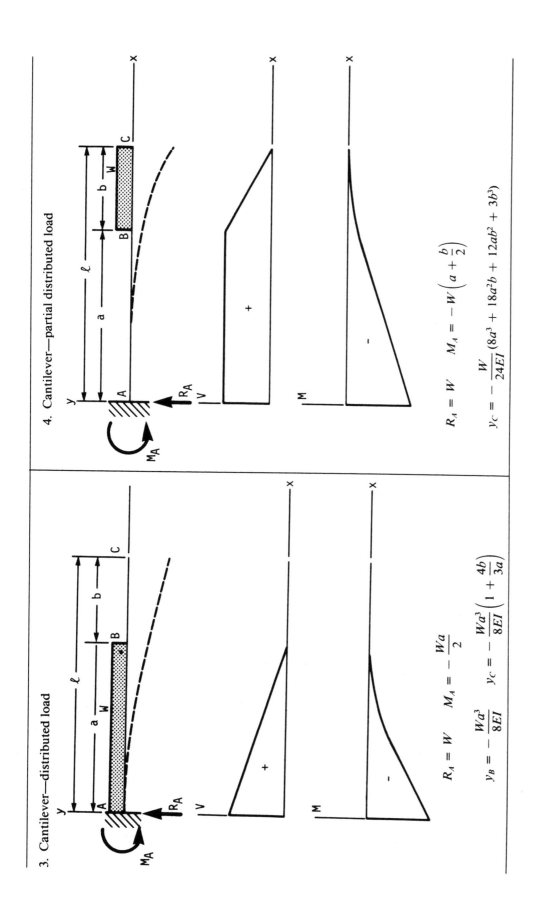

3. Cantilever—distributed load

$$R_A = W \qquad M_A = -\frac{Wa}{2}$$

$$y_B = -\frac{Wa^3}{8EI} \qquad y_C = -\frac{Wa^3}{8EI}\left(1 + \frac{4b}{3a}\right)$$

4. Cantilever—partial distributed load

$$R_A = W \qquad M_A = -W\left(a + \frac{b}{2}\right)$$

$$y_C = -\frac{W}{24EI}(8a^3 + 18a^2b + 12ab^2 + 3b^3)$$

79

TABLE 3-1 Properties of Beams (*Continued*)

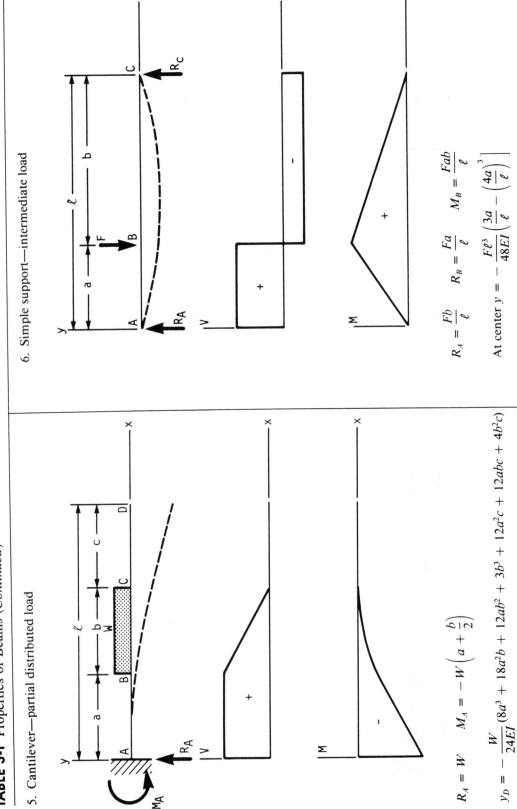

6. Simple support—intermediate load

$$R_A = \frac{Fb}{\ell} \qquad R_B = \frac{Fa}{\ell} \qquad M_B = \frac{Fab}{\ell}$$

$$\text{At center } y = -\frac{F\ell^3}{48EI}\left(\frac{3a}{\ell} - \left(\frac{4a}{\ell}\right)^3\right)$$

5. Cantilever—partial distributed load

$$R_A = W \qquad M_A = -W\left(a + \frac{b}{2}\right)$$

$$y_D = -\frac{W}{24EI}(8a^3 + 18a^2b + 12ab^2 + 3b^3 + 12a^2c + 12abc + 4b^2c)$$

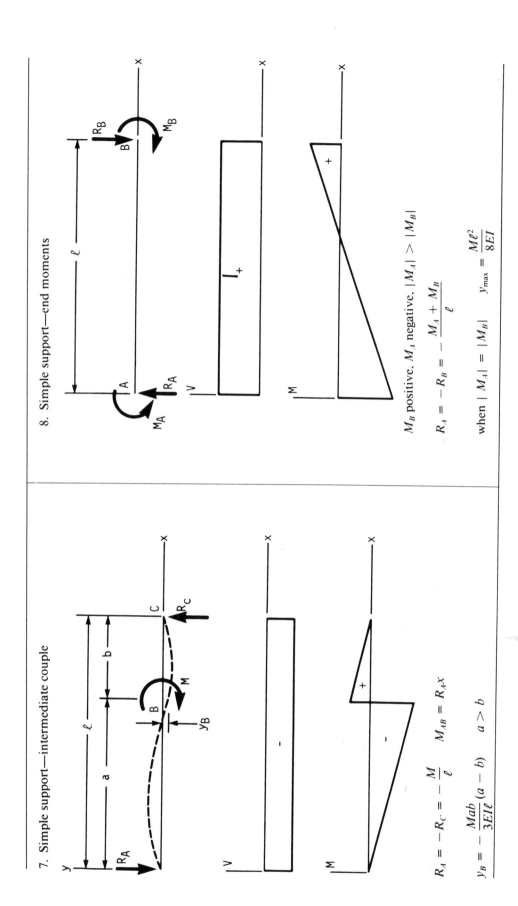

7. Simple support—intermediate couple

$$R_A = -R_C = -\frac{M}{\ell} \qquad M_{AB} = R_A x \qquad a > b$$

$$y_B = -\frac{Mab}{3EI\ell}(a - b)$$

8. Simple support—end moments

M_B positive, M_A negative, $|M_A| > |M_B|$

$$R_A = -R_B = -\frac{M_A + M_B}{\ell}$$

when $|M_A| = |M_B| \qquad y_{max} = \frac{M\ell^2}{8EI}$

TABLE 3-1 Properties of Beams (*Continued*)

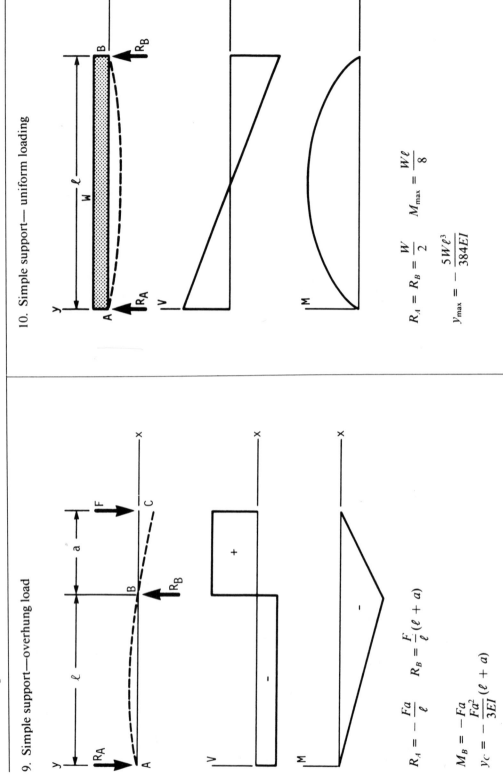

9. Simple support—overhung load

$$R_A = -\frac{Fa}{\ell} \qquad R_B = \frac{F}{\ell}(\ell + a)$$

$$M_B = -Fa$$

$$y_C = -\frac{Fa^2}{3EI}(\ell + a)$$

10. Simple support— uniform loading

$$R_A = R_B = \frac{W}{2} \qquad M_{max} = \frac{W\ell}{8}$$

$$y_{max} = -\frac{5W\ell^3}{384EI}$$

82

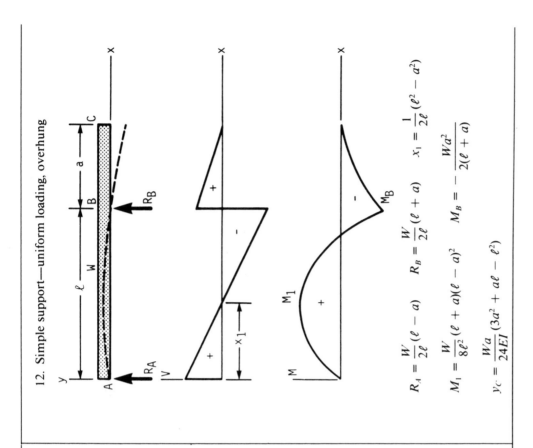

12. Simple support—uniform loading, overhung

$$R_A = \frac{W}{2\ell}(\ell - a) \qquad R_B = \frac{W}{2\ell}(\ell + a) \qquad x_1 = \frac{1}{2\ell}(\ell^2 - a^2)$$

$$M_1 = \frac{W}{8\ell^2}(\ell + a)(\ell - a)^2 \qquad M_B = -\frac{Wa^2}{2(\ell + a)}$$

$$y_C = \frac{Wa}{24EI}(3a^2 + a\ell - \ell^2)$$

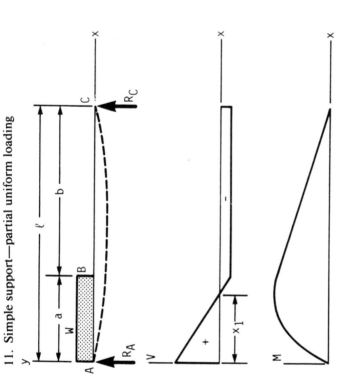

11. Simple support—partial uniform loading

$$R_A = \frac{W}{2\ell}(2\ell - a) \qquad R_B = \frac{Wa}{2\ell} \qquad x_1 = \frac{a}{2\ell}(2\ell - a) \qquad a < \frac{\ell}{2}$$

At center $y = -\dfrac{Wa}{48EI}(a^2 + 2\ell^2)$

83

TABLE 3-1 Properties of Beams (*Continued*)

13. Simple support—overhung uniform load

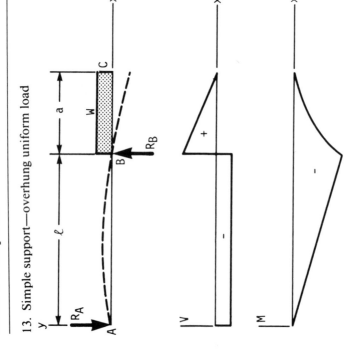

$$R_A = -\frac{Wa}{2\ell} \qquad R_B = \frac{W}{2\ell}(2\ell + a) \qquad M_{max} = -\frac{Wa}{2}$$

$$y_{max} = \frac{0.032\,Wa\ell^2}{EI} \qquad \text{between supports}$$

$$y_C = -\frac{Wa^2}{24EI}(4\ell + 3a)$$

14. Fixed and simple support—intermediate load

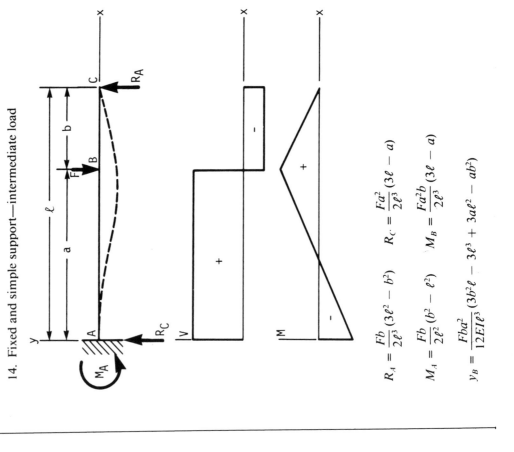

$$R_A = \frac{Fb}{2\ell^3}(3\ell^2 - b^2) \qquad R_C = \frac{Fa^2}{2\ell^3}(3\ell - a)$$

$$M_A = \frac{Fb}{2\ell^2}(b^2 - \ell^2) \qquad M_B = \frac{Fa^2b}{2\ell^3}(3\ell - a)$$

$$y_B = \frac{Fba^2}{12EI\ell^3}(3b^2\ell - 3\ell^3 + 3a\ell^2 - ab^2)$$

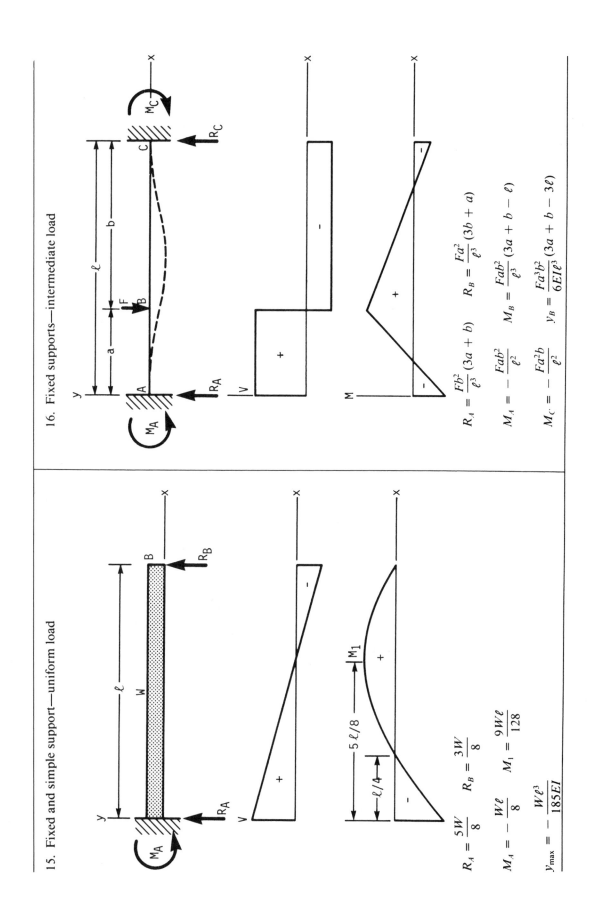

15. Fixed and simple support—uniform load

$$R_A = \frac{5W}{8} \qquad R_B = \frac{3W}{8}$$

$$M_A = -\frac{W\ell}{8} \qquad M_1 = \frac{9W\ell}{128}$$

$$y_{\max} = -\frac{W\ell^3}{185EI}$$

16. Fixed supports—intermediate load

$$R_A = \frac{Fb^2}{\ell^3}(3a+b) \qquad R_B = \frac{Fa^2}{\ell^3}(3b+a)$$

$$M_A = -\frac{Fab^2}{\ell^2} \qquad M_B = \frac{Fab^2}{\ell^3}(3a+b-\ell)$$

$$M_C = -\frac{Fa^2b}{\ell^2} \qquad y_B = \frac{Fa^3b^2}{6EI\ell^3}(3a+b-3\ell)$$

TABLE 3-1 Properties of Beams (*Continued*)

17. Fixed supports—uniform load

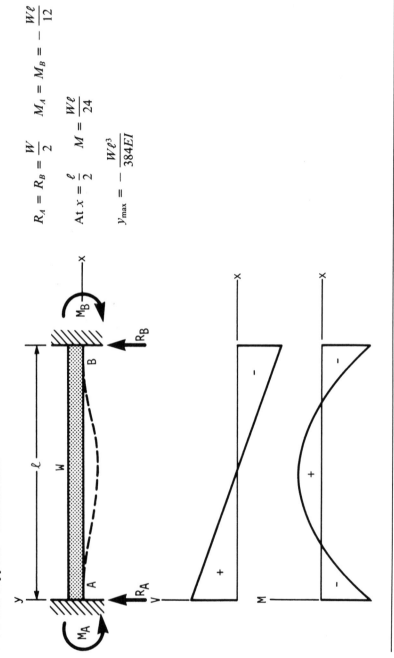

$$R_A = R_B = \frac{W}{2} \qquad M_A = M_B = -\frac{W\ell}{12}$$

$$\text{At } x = \frac{\ell}{2} \qquad M = \frac{W\ell}{24}$$

$$y_{max} = -\frac{W\ell^3}{384EI}$$

In using Table 3-1, remember that the deflection at the center of a beam with off-center loads is usually within 2.5 percent of the maximum value.

3-4 COMPUTER ANALYSIS

In this section we will develop a computer method using numerical analysis to determine the slope and deflection of any simply supported beam having a variety of concentrated loads, including point couples, with any number of step changes in cross section. The method is particularly applicable to stepped shafts where the transverse bending deflections and neutral-axis slopes are desired at specified points.

The method uses numerical analysis to integrate Eq. (3-6) twice in a marching method. The first integration uses the trapezoidal rule; the second uses Simpson's rule. The procedure gives exact results.

Let us define the two successive integrals as

$$\phi = \int_0^x \frac{M}{EI} \, dx \qquad \psi = \int_0^x \phi \, dx \tag{3-9}$$

But from Eq. (3-7), the slope is

$$\theta = \frac{dy}{dx} = \int_0^x \frac{M}{EI} \, dx + C_1$$
$$= \phi + C_1 \tag{a}$$

A second integration gives

$$y = \psi + C_1 x + C_2 \tag{b}$$

It is convenient to write Eqs. (a) and (b) as

$$\theta = K(\phi + C_1) \tag{3-10}$$

$$y = K(\psi + C_1 x + C_2) \tag{3-11}$$

where K depends on the units used.

Locating supports at $x = a$ and $x = b$ and specifying zero deflection at these supports provides the two conditions for finding C_1 and C_2. The results are

$$C_1 = \frac{\psi_b - \psi_a}{x_a - x_b} \tag{3-12}$$

$$C_2 = \frac{x_b \psi_a - x_a \psi_b}{x_a - x_b} \tag{3-13}$$

Now we write the first of Eqs. (3-9) using the trapezoidal rule:

$$\phi_{i+2} = \phi_i + \frac{1}{2} \left[\left(\frac{M}{EI} \right)_{i+1} + \left(\frac{M}{EI} \right)_i \right] (x_{i+2} - x_i) \tag{3-14}$$

Applying Simpson's rule to the second of Eqs. (3-9) yields

$$\psi_{i+4} = \psi_i + \tfrac{1}{6}(\phi_{i+4} + 4\phi_{i+2} + \phi_i)(x_{i+4} - x_i) \tag{3-15}$$

Notes ▪ Drawings ▪ Ideas

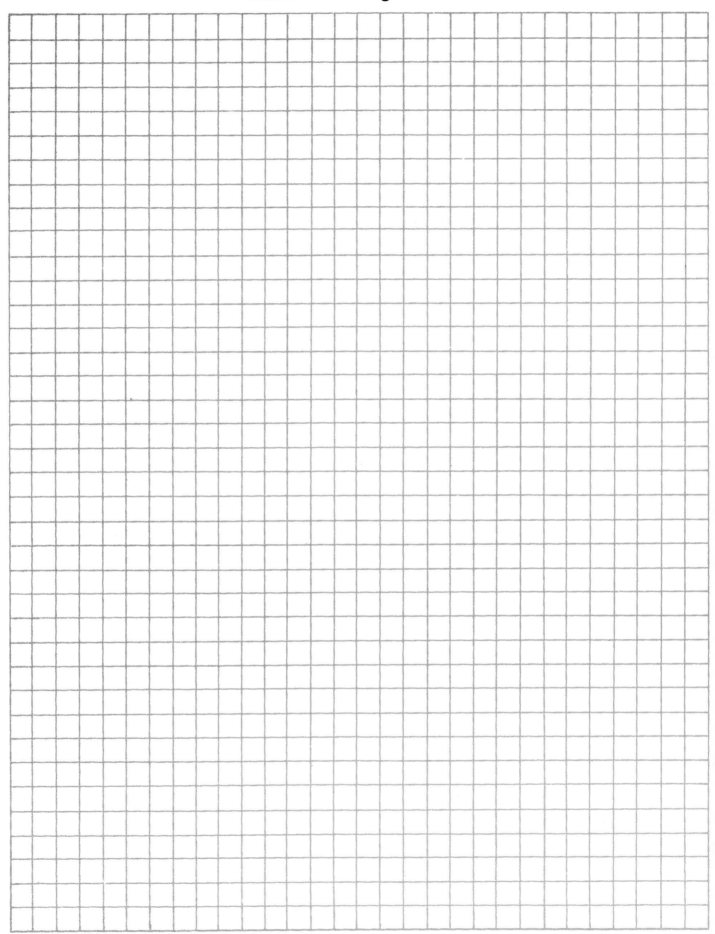

As indicated previously, these equations are used in a marching manner. Thus, using Eq. (3-14) we successively compute ϕ_1, ϕ_3, ϕ_5, . . . beginning at x_1 and ending at x_N, where N is the number of M/EI values. Similarly, Eq. (3-15) is integrated successively to yield ψ_1, ψ_5, ψ_9, . . . , ψ_N.

After these two integrations have been performed, the constants C_1 and C_2 can be found from Eqs. (3-12) and (3-13), and then Eqs. (3-10) and (3-11) can be solved for the deflection and slope. These terms will have the same indices as the integral ψ.

The details of the method are best explained by an example. The shaft of Fig. 3-2 has all points of interest designated by the station letters A, B, C, ... These points must include

- Location of all supports and concentrated loads
- Location of cross-sectional changes
- Location of points at which the deflection and slope are desired

Refer now to Table 3-2 and note that coordinates x tabulated in column 2 correspond to each station. Note also the presence of additional x coordinates; these are selected as halfway stations.

Column 4 of Table 3-2 shows that two M/EI values must be computed for each x coordinate. These are needed to account for the fact that M/EI has an abrupt change at every shoulder or change in cross section.

The indices $i = 1, 2, 3, . . . , N$ in Eqs. (3-14) and (3-15) correspond to the M/EI values and are shown in column 3 of Table 3-2. A program in BASIC is shown in Fig. 3-3. Note that the term M(I) is used for $(M/EI)_i$.

3-5 ANALYSIS OF FRAMES

Castigliano's theorem is introduced in Chap. 5, and the energy equations needed for its use are listed in Table 5-2. The method can be used to find the deflection at any point of a frame such as the one shown in Fig. 3-4. For example, the deflection δ_C at C in the direction of F_2 can be found using Eq. (5-2) as

$$\delta_C = \frac{\partial U}{\partial F_2} \qquad (a)$$

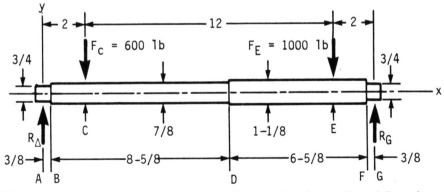

FIG. 3-2 Simply supported stepped shaft loaded by forces F_C and F_E and supported by bearing reactions R_A and R_G. All dimensions in inches.

TABLE 3-2 Summary of Beam Computations†

Station (1)	x (2)	N (3)	M/EI (4)	φ (5)	ψ (6)	y (7)	θ (8)
A	0	0 1	0 0	0	0	0	−1.028E-02
	0.188	2 3	261.6 261.6	24.59			
B	0.375	4 5	523.2 282.4	98.0	12.27	−3.8444E-03	−1.019E-02
	1.188	6 7	894.6 894.6	576.4			
C	2	8 9	1 506 1 506	1 551	1 083	−0.0195	−8.733E-03
	5.5	10 11	1 708.7 1 708.7	7 177			
D	9	12 13	1 911.4 699.5	13 512	52 149	−4.0408E-02	3.228E-03
	11.5	14 15	752.5 752.5	15 327			
E	14	16 17	805.5 805.5	17 274	128 894	−1.5084E-02	6.990E-03
	14.813	18 19	478.1 478.1	17 796			
F	15.625	20 21	151.0 764.7	18 052	157 741	−2.9488E-03	7.768E-03
	15.813	22 23	382.3 382.3	18 159			
G	16	24 25	0 0	18 195	164 546	0	7.911E-03

†The units are in for x, lb·in for M, Mpsi for E, in^4 for I, in for y, and rad for θ.

where U = the strain energy stored in the entire frame due to all the forces. If the deflection is desired in another direction or at a point where no force is acting, then a fictitious force Q is added to the system at that point and in the direction in which the deflection is desired. After the partial derivatives have been found, Q is equated to zero, and the remaining terms give the wanted deflection.

```
10 PRINT "YOU MAY USE EITHER U.S. CUSTOMARY UNITS IN THIS PROGRAM"
20 PRINT "OR METRIC UNITS.  IF U.S. CUSTOMARY UNITS ARE USED"
30 PRINT "M IS IN INCH-POUNDS, E IN MPSI, AND I IN INCHES TO"
40 PRINT "THE FOURTH POWER.  IF METRIC UNITS ARE USED, M IS IN"
50 PRINT "NEWTON-METERS, E IN GPA, AND I IN CENTIMETERS TO THE"
60 PRINT "FOURTH POWER."
70 PRINT "WILL YOU USE METRIC UNITS (Y OR N)";U$
80 INPUT U$
90 IF U$= "Y" THEN 100 ELSE 110
100 K = .0001:GOTO 120
110 K = .000001
120 DIM M(65),X(65),PHI(65),PSI(65),Y(65),THETA(65)
130 INPUT "N=";N: FOR I = 1 TO N
140 INPUT "M=";M(I): LPRINT "M("I")="M(I)
150 NEXT I
160 FOR I = 1 TO N STEP 2
170 INPUT "X=";X(I): LPRINT "X("I")="X(I)
180 NEXT I
190 STOP
200 LPRINT "PHI( 1 )="PHI(1)
210 FOR I = 1 TO (N-2) STEP 2
220 PHI(I+2) = PHI(I) + ((M(I+1) + M(I))*(X(I+2) - X(I))*.5)
230 LPRINT "PHI("I+2")="PHI(I+2)
240 NEXT I
250 LPRINT "PSI( 1 )="PSI(1)
260 FOR I = 1 TO (N-4) STEP 4
270 PSI(I+4)=PSI(I)+(((PHI(I+4)+(4*PHI(I+2))+PHI(I))*(X(I+4)-X(I)))/6)
280 LPRINT "PSI("I+4")="PSI(I+4)
290 NEXT I
300 PRINT "SPECIFY VALUES OF X AND PSI AT SUPPORT A"
310 INPUT "X=";A    : LPRINT "X(A)="A
320 INPUT "PSI=";PSIA  : LPRINT "PSI(A)="PSIA
330 LPRINT
340 PRINT "SPECIFY VALUES OF X AND PSI AT SUPPORT B"
350 INPUT "X=";B    : LPRINT "X(B)="B
360 INPUT "PSI=";PSIB  : LPRINT "PSI(B)="PSIB
370 LPRINT
380 C1  = (PSIB - PSIA)/(A-B)
390 LPRINT "C(1)="C1
400 C2 =((B*PSIA)-(A*PSIB))/(A-B)
410 LPRINT "C(2)="C2
420 LPRINT
430 FOR I = 1 TO N STEP 4
440 IF X(I) = A THEN 450 ELSE 460
450 Y(I)=0: GOTO 490
460 IF X(I)=B THEN 470 ELSE 480
470 Y(I) = 0: GOTO 490
480 Y(I) =(PSI(I) + (C1  *X(I)) + C2)*K
490 LPRINT "Y("I")="Y(I)
500 NEXT I
510 LPRINT
520 FOR I = 1 TO N STEP 4
530 THETA(I) =(PHI(I) + C1)*K
540 LPRINT "THETA("I")="THETA(I)
550 NEXT I
560 END
```

FIG. 3-3 Beam problem programmed in BASIC.

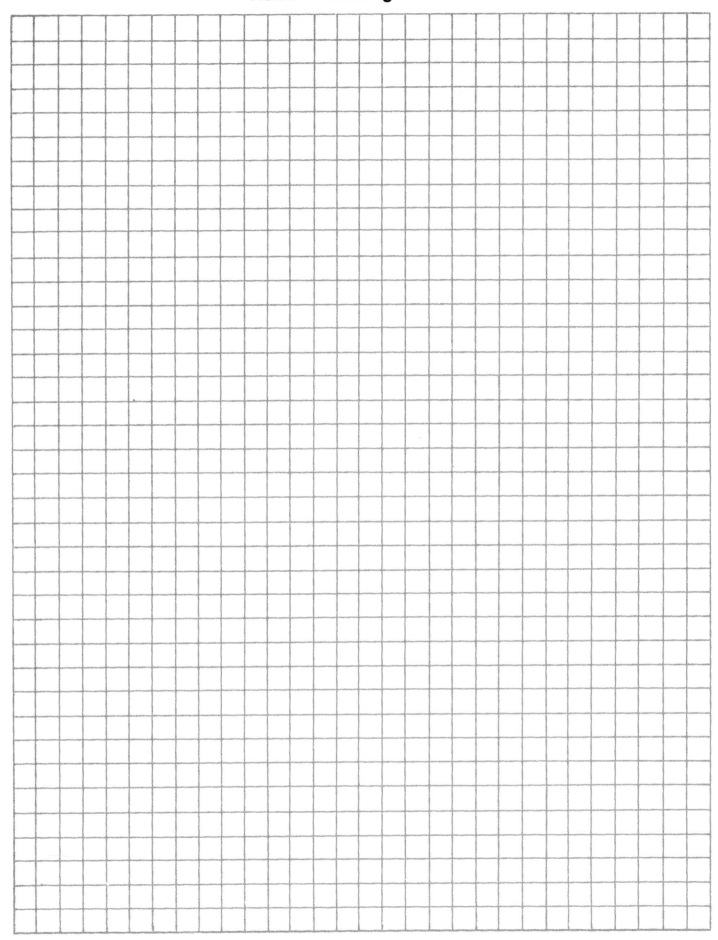

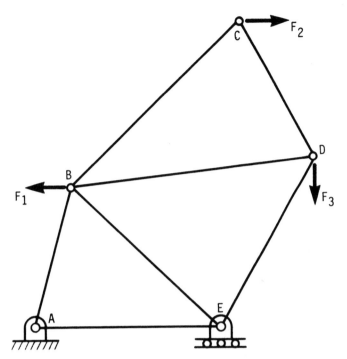

FIG. 3-4   Frame loaded by three forces.

The first step in using the method is to make a force analysis of each member of the frame. If Eq. (*a*) is to be solved, then the numerical values of F_1 and F_2 can be used in the force analysis, but the value of F_2 must *not* be substituted until after each member has been analyzed and the partial derivatives obtained. The following example demonstrates the technique.

EXAMPLE 1. Find the downward deflection of point D of the frame shown in Fig. 3-5.

Solution. A force analysis of the system gives an upward reaction at E of $R_E = 225 + 3F_2$. The reaction at A is downward and is $R_A = 75 - 2F_2$.

The strain energy for member CE is

$$U_{CE} = \frac{R_A^2 \ell}{2AE} \tag{1}$$

The partial deflection is taken with respect to F_2 because the deflection at D in the direction of F_2 is desired. Thus

$$\frac{\partial U_{CE}}{\partial F_2} = \frac{2R_A\ell}{2AE} \frac{\partial R_A}{\partial F_2} \tag{2}$$

Also,

$$\frac{\partial R_A}{\partial F_2} = -2$$

Thus Eq. (2) becomes

$$\frac{\partial U_{CE}}{\partial F_2} = \frac{(75 - 2F_2)(30)}{0.2E}(-2) = \frac{37\,500}{E} \tag{3}$$

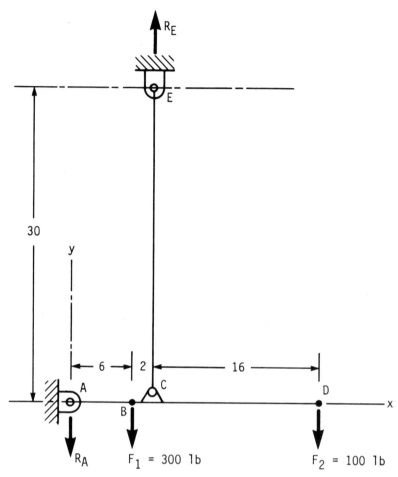

FIG. 3-5 Frame loaded by two forces. Dimensions in inches: $A_{CE} = 0.20$ in²; $I_{AD} = 0.18$ in⁴; $E = 30 \times 10^6$ psi.

Note that we were able to substitute the value of F_2 in Eq. (3) because the partial derivative had been taken.

The strain energy stored in member $ABCD$ will have to be computed in three parts because of the change in direction of the bending moment diagram at points B and C. For part AB, the moment is

$$M_{AB} = R_A x = (75 - 2F_2)x$$

The strain energy is

$$U_{AB} = \int_0^6 \frac{M_{AB}^2}{2EI} \, dx \tag{4}$$

Taking the partial derivative with respect to F_2 as before gives

$$\frac{\partial U_{AB}}{\partial F_2} = \int_0^6 \frac{2M_{AB}}{2EI} \frac{\partial M_{AB}}{\partial F_2} \, dx \tag{5}$$

But

$$\frac{\partial M_{AB}}{\partial F_2} = -2x \tag{6}$$

Therefore, Eq. (5) may be written

$$\frac{\partial U_{AB}}{\partial F_2} = \frac{1}{EI} \int_0^6 x(75 - 2F_2)(-2x)\,dx \tag{7}$$

$$= \frac{1}{0.18E} \int_0^6 250x^2\,dx = \frac{100\,000}{E}$$

where the value of F_2 again has been substituted after taking the partial derivative.
   For section BC, we have

$$M_{BC} = R_A x - F_1(x - 6) = 1800 - 225x - 2F_2 x$$

$$\frac{\partial M_{BC}}{\partial F_2} = -2x$$

$$\frac{\partial U_{BC}}{\partial F_2} = \int_6^8 \frac{2M_{BC}}{2EI}\frac{\partial M_{BC}}{\partial F_2}\,dx$$

$$= \frac{1}{EI} \int_6^8 (1800 - 225x - 2F_2 x)(-2x)\,dx$$

$$= \frac{1}{0.18E} \int_6^8 (-3600x + 850x^2)\,dx = \frac{145\,926}{E}$$

Finally, section CD yields

$$M_{CD} = -(24 - x)F_2 \qquad \frac{\partial M_{CD}}{\partial F_2} = -(24 - x)$$

$$\frac{\partial U_{CD}}{\partial F_2} = \int_8^{24} \frac{2M_{CD}}{2EI}\frac{\partial M_{CD}}{\partial F_2}\,dx$$

$$= \frac{1}{EI} \int_8^{24} F_2(24 - x)^2\,dx$$

$$= \frac{1}{0.18E} \int_8^{24} (57\,600 - 4800x + 100x^2)\,dx$$

$$= \frac{758\,519}{E}$$

Then

$$y_D = \frac{\partial U_{CE}}{\partial F_2} + \frac{\partial U_{AB}}{\partial F_2} + \frac{\partial U_{BC}}{\partial F_2} + \frac{\partial U_{CD}}{\partial F_2}$$

$$= \frac{1}{30(10)^6}(37\,500 + 100\,000 + 145\,926 + 758\,519)$$

$$= 0.0347 \text{ in} \quad \text{(when rounded)}$$

3-5-1 Redundant Members

A frame consisting of one or more redundant members is statically indeterminate because the use of statics is not sufficient to determine all the reactions. In this case, Castigliano's theorem can be used first to determine these reactions and second to determine the desired deflection.

Let R_1, R_2, and R_3 be a set of three indeterminate reactions. The deflection at the supports must be zero, and so Castigliano's theorem can be written three times. Thus

$$\frac{\partial U}{\partial R_1} = 0 \qquad \frac{\partial U}{\partial R_2} = 0 \qquad \frac{\partial U}{\partial R_3} = 0 \qquad (b)$$

and so the number of equations to be solved is the same as the number of indeterminate reactions.

In setting up Eqs. (b), *do not* substitute the numerical value of the particular force corresponding to the desired deflection. This force symbol must appear in the reaction equations because the partial derivatives must be taken with respect to this force when the deflection is found. The method is illustrated by the following example.

EXAMPLE 2. Find the downward deflection at point D of the frame shown in Fig. 3-6.

Solution. Choose R_B as the statically indeterminate reaction. A static force analysis then gives the remaining reactions as

$$R_A = R_C = 0.625(F - R_B) \qquad (1)$$

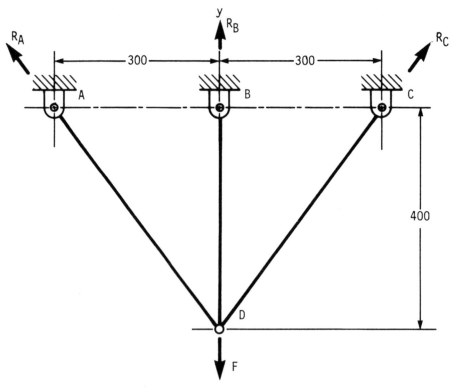

FIG. 3-6 Frame loaded by a single force. Dimensions in millimeters: $A_{AD} = A_{CD} = 2 \text{ cm}^2$, $A_{BD} = 1.2 \text{ cm}^2$, $E = 207 \text{ GPa}$, $F = 20 \text{ kN}$.

The frame consists only of tension members, so the strain energy in each member is

$$U_{AD} = U_{DC} = \frac{R_A^2 \ell_{AD}}{2A_{AD}E} \qquad U_{BD} = \frac{R_B^2 \ell_{BD}}{2A_{BD}E} \tag{2}$$

Using Eq. (b), we now write

$$0 = \frac{\partial U}{\partial R_B} = \frac{2R_A \ell_{AD}}{A_{AD}E} \frac{\partial R_A}{\partial R_B} + \frac{R_B \ell_{BD}}{A_{BD}E} \frac{\partial R_B}{\partial R_B} \tag{3}$$

Equation (1) gives $\partial R_A / \partial R_B = -0.625$. Also, $\partial R_B / \partial R_B = 1$. Substituting numerical values in Eq. (3), except for F, gives

$$\frac{2(0.625)(F - R_B)(500)(-0.625)}{2(207)} + \frac{R_B(400)(1)}{1.2(207)} = 0 \tag{4}$$

Solving gives $R_B = 0.369F$. Therefore, from Eq. (1), $R_A = R_C = 0.394F$. This completes the solution of the case of the redundant member. The next problem is to find the deflection at D.

Using Eq. (2), again we write

$$y_D = \frac{\partial U}{\partial F} = \frac{2R_A \ell_{AD}}{A_{AD}E} \frac{\partial R_A}{\partial F} + \frac{R_B \ell_{BD}}{A_{BD}E} \frac{\partial R_B}{\partial F} \tag{5}$$

For use in this equation we note that $\partial R_A / \partial F = 0.394$ and $\partial R_B / \partial F = 0.369$. Having taken the derivatives, we can now substitute the numerical value of F. Thus Eq. (5) becomes†

$$y_D = \left\{ \frac{2[0.394(20)](500)(0.394)}{2(207)} + \frac{[0.369(20)](400)(0.369)}{1.2(207)} \right\} 10^{-2}$$

$$= 0.119 \text{ mm}$$

---

†In general, when using metric quantities, prefixed units are chosen so as to produce number strings of not more than four members. Thus some preferred units in SI are MPa (N/mm²) for stress, GPa for modulus of elasticity, mm for length, and, say, cm⁴ for second moment of area.

People are sometimes confused when they encounter an equation containing a number of mixed units. Suppose we wish to solve a deflection equation of the form

$$y = \frac{64F\ell^3}{3\pi d^4 E}$$

where $F = 1.30$ kN, $\ell = 300$ mm, $d = 2.5$ cm, and $E = 207$ GPa. Form the equation into two parts, the first containing the numbers and the second containing the prefixes. This converts everything to base units including the result. Thus,

$$y = \frac{64(1.30)(300)^3}{3\pi(2.5)^4(207)} \frac{(\text{kilo})(\text{milli})^3}{(\text{centi})^4(\text{giga})}$$

Now compute the numerical value of the first part and substitute the prefix values in the second. This gives

$$y = (29.48 \times 10^3) \left[\frac{10^3(10^{-3})^3}{(10^{-2})^4(10^9)} \right] = 29.48 \times 10^{-4} \text{ m}$$

$$= 2.948 \text{ mm}$$

Note that we multiplied the result by 10^3 mm/m to get the answer in millimeters. When this approach is used with Eq. (5) it is found that the result must be multiplied by $(10)^{-2}$ to get y in millimeters.

Notes • Drawings • Ideas

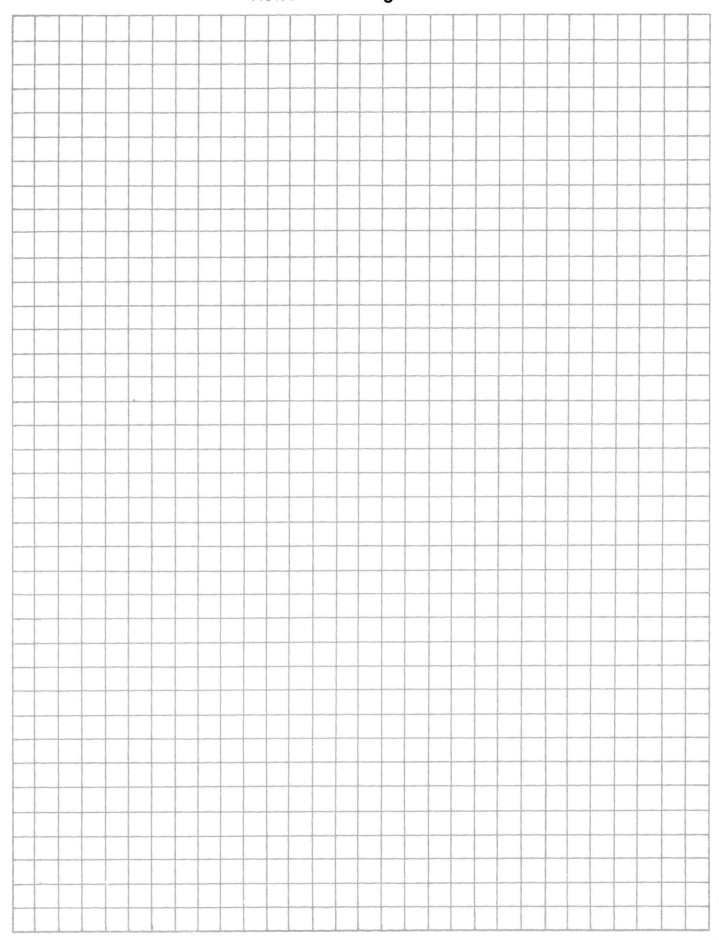

Notes • Drawings • Ideas

chapter 4
INSTABILITIES IN BEAMS AND COLUMNS

HARRY HERMAN
Professor of Mechanical Engineering
New Jersey Institute of Technology
Newark, New Jersey

NOTATION

A	Area of cross section
$B(n)$	Arbitrary constants
$c(n)$	Coefficients in series
$c(y)$, $c(z)$	Distance from y and z axis, respectively to outermost compressive fiber
e	Eccentricity of axial load P
E	Modulus of elasticity of material
$E(t)$	Tangent modulus for buckling outside of elastic range
$F(x)$	A function of x
G	Shear modulus of material
h	Height of cross section
H	Horizontal (transverse) force on column
I	Moment of inertia of cross section
$I(y)$, $I(z)$	Moments of inertia with respect to y and z axis, respectively
J	Torsion constant; polar moment of inertia
k^2	P/EI
K	Effective-length coefficient
$K(0)$	Spring constant for constraining spring at origin
$K(T, 0)$, $K(T, L)$	Torsional spring constants at $x = 0$, L, respectively
l	Developed length of cross section
L	Length of column or beam
L_{eff}	Effective length of column
M, M'	Bending moments
$M(0)$, $M(L)$, M_{mid}	Bending moments at $x = 0$, L, and midpoint, respectively
$M(0)_{\text{cr}}$	Critical moment for buckling of beam
M_{tr}	Moment due to transverse load

$M(y)$, $M(z)$	Moment about y and z axis, respectively
n	Integer; running index
P	Axial load on column
P_{cr}	Critical axial load for buckling of column
r	Radius of gyration
R	Radius of cross section
s	Running coordinate, measured from one end
t	Thickness of cross section
T	Torque about x axis
x	Axial coordinate of column or beam
y, z	Transverse coordinates and deflections
Y	Initial deflection (crookedness) of column
Y_{tr}	Deflection of beam-column due to transverse load
η	Factor of safety
σ	Stress
ϕ	Angle of twist

As the terms *beam* and *column* imply, this chapter deals with members whose cross-sectional dimensions are small in comparison with their lengths. Particularly, we are concerned with the stability of beams and columns whose axes in the undeformed state are substantially straight. Classically, instability is associated with a state in which the deformation of an idealized, perfectly straight member can become arbitrarily large. However, some of the criteria for stable design which we will develop will take into account the influences of imperfections such as the eccentricity of the axial load and crookedness of the centroidal axis of the column. The magnitudes of these imperfections are generally not known, but they can be estimated from manufacturing tolerances. For axially loaded columns, the onset of instability is related to the moment of inertia of the column cross section about its minor principal axis. For beams, stability design requires, in addition to the moment of inertia, the consideration of the torsional stiffness.

4-1 EULER'S FORMULA

We will begin with the familiar Euler column-buckling problem. The column is idealized as shown in Fig. 4-1. The top and bottom ends are pinned; that is, the moments at the ends are zero. The bottom pin is fixed against translation; the top pin is free to move in the vertical direction only; the force P acts along the x axis, which coincides with the centroidal axis in the undeformed state. It is important to keep in mind that the analysis which follows applies only to columns with cross sections and loads that are symmetrical about the xy plane in Fig. 4-1 and satisfy the usual assumptions of linear beam theory. It is particularly important in this connection to keep in mind that this analysis is valid only when the deformation is such that the

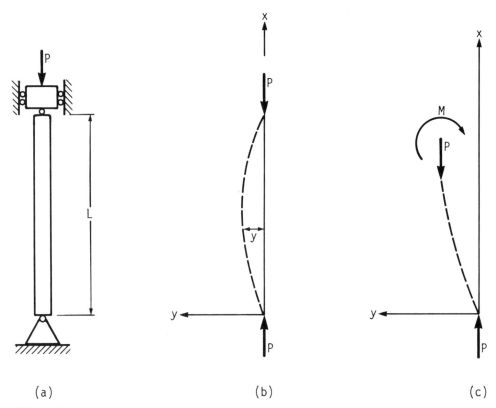

FIG. 4-1 Deflection of a simply supported column. (*a*) Ideal simply supported column; (*b*) column-deflection curve; (*c*) free-body diagram of deflected segment.

square of the slope of the tangent at any point on the deflection curve is negligibly small compared to unity (fortunately, this is generally true in design applications). In such a case, the familiar differential equation for the bending of a beam is applicable. Thus,

$$EI\frac{d^2y}{dx^2} = M \tag{4-1}$$

For the column in Fig. 4-1,

$$M = -Py \tag{4-2}$$

We take E and I as constant, and let

$$\frac{P}{EI} = k^2 \tag{4-3}$$

Then we get, from Eqs. (4-1), (4-2), and (4-3),

$$\frac{d^2y}{dx^2} + k^2y = 0 \tag{4-4}$$

The boundary conditions at $x = 0$ and $x = L$ are

$$y(0) = y(L) = 0 \tag{4-5}$$

In order that Eqs. (4-4) and (4-5) should have a solution $y(x)$ that need not be equal to zero for all values of x, k must take one of the values in Eq. (4-6):

$$k(n) = \frac{n\pi}{L} \qquad n = 1, 2, 3, \ldots \qquad (4\text{-}6)$$

which means that the axial load P must take one of the values in Eq. (4-7):

$$P(n) = \frac{n^2\pi^2 EI}{L^2} \qquad n = 1, 2, 3, \ldots \qquad (4\text{-}7)$$

For each value of n, the corresponding nonzero solution for y is

$$y(n) = B(n) \sin\left(\frac{n\pi x}{L}\right) \qquad n = 1, 2, 3, \ldots \qquad (4\text{-}8)$$

where $B(n)$ is an arbitrary constant.

In words, the preceding results state the following: Suppose that we have a perfectly straight prismatic column with constant properties over its entire length. If the column is subjected to a perfectly axial load, there is a set of load values, together with a set of sine-shaped deformation curves for the column axis, such that the applied moment due to the axial load and the resisting internal moment are in equilibrium everywhere along the column no matter what the amplitude of the sine curve may be. From Eq. (4-7), the smallest load at which such deformation occurs, called the *critical load,* is

$$P_{\text{cr}} = \frac{\pi^2 EI}{L^2} \qquad (4\text{-}9)$$

This is the familiar Euler formula.

4-2 EFFECTIVE LENGTH

Note that the sinusoidal shape of the solution function is determined by the differential equation and does not depend on the boundary conditions. If we can find a segment of a sinusoidal curve that satisfies our chosen boundary conditions and, in turn, we can find some segment of that curve which matches the curve in Fig. 4-1, we can establish a correlation between the two cases. This notion is the basis for the "effective-length" concept. Recall that Eq. (4-9) was obtained for a column with both ends simply supported (that is, the moment is zero at the ends). Figure 4-2 illustrates columns of length L with various idealized end conditions. In each case, there is a multiple of L, KL, which is called the *effective length of the column* L_{eff}, that has a shape which is similar to and behaves like a simply supported column of that length. To determine the critical loads for columns whose end supports may be idealized as shown in Fig. 4-2, we can make use of Eq. (4-9) if we replace L by KL, with the appropriate value of K taken from Fig. 4-2. Particular care has to be taken to distinguish between the case in Fig. 4-2c, where both ends of the column are secured against rotation and transverse translation, versus the case in Fig. 4-2e, where the ends do not rotate, but relative transverse movement of one end of the column with respect to the other end is possible. The effective length in the first case is half that in the second case, so that the critical load in the first case is four times that in the second case. A major difficulty with using the results in Fig. 4-2 is that

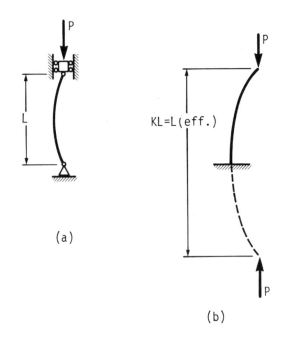

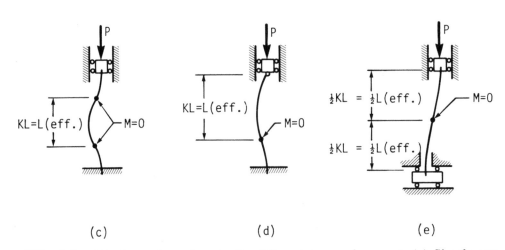

FIG. 4-2 Effective column lengths for different types of support. (*a*) Simply supported, $K = 1$; (*b*) fixed-free, $K = 2$; (*c*) fixed-fixed, $K = \frac{1}{2}$; (*d*) fixed-pinned, $K = 0.707$; (*e*) ends nonrotating, but have transverse translation.

in real problems a column end is seldom perfectly fixed or perfectly free (even approximately) with regard to translation or rotation. In addition, we must remember that the critical load is inversely proportional to the square of the effective length. Thus a change of 10 percent in L_{eff} will result in a change of about 20 percent in the critical load, so that a fair approximation of the effective length produces an unsatisfactory approximation of the critical load. We will now develop more general results that will allow us to take into account the elasticity of the structure surrounding the column.

4-3   GENERALIZATION OF THE PROBLEM

We will begin with a generalization of the case in Fig. 4-2e. In Fig. 4-3, the lower end is no longer free to translate, but instead is elastically constrained. The differential equation is

$$EI \frac{d^2y}{dx^2} = M = M(0) + P[y(0) - y] + Hx \tag{4-10}$$

Here H, the horizontal force at the origin, may be expressed in terms of the deflection at the origin $y(0)$ and the constant of the constraining spring $K(0)$:

$$H = -K(0)y(0) \tag{4-11}$$

$M(0)$ is the moment which prevents rotation of the beam at the origin. The moment which prevents rotation of the beam at the end $x = L$ is $M(L)$. The boundary conditions are

$$y(L) = 0 \qquad \frac{dy(L)}{dx} = 0 \qquad \frac{dy(0)}{dx} = 0 \tag{4-12}$$

The rest of the symbols are the same as before. We define k as in Eq. (4-3). As in the case of the simply supported column, Eqs. (4-10), (4-11), and (4-12) have solutions in which $y(x)$ need not be zero for all values of x, but again these solutions occur only for certain values of kL. Here these values of kL must satisfy Eq. (4-13):

$$[2(1 - \cos kL) - kL \sin kL]L^3 K(0) + EI(kL)^3 \sin kL = 0 \tag{4-13}$$

The physical interpretation is the same as in the simply supported case. If we denote the lowest value of kL that satisfies Eq. (4-13) by $(kL)_{cr}$, then the column buckling load is given by

$$P_{cr} = \frac{EI(kL)_{cr}^2}{L^2} \tag{4-14}$$

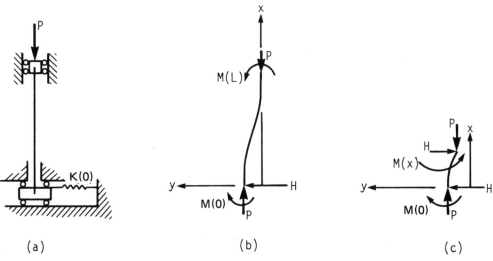

(a)   (b)   (c)

FIG. 4-3   Column with ends fixed against rotation and an elastic end constraint against transverse deflection. (a) Undeflected column; (b) deflection curve; (c) free-body diagram of deflected segment.

Notes · Drawings · Ideas

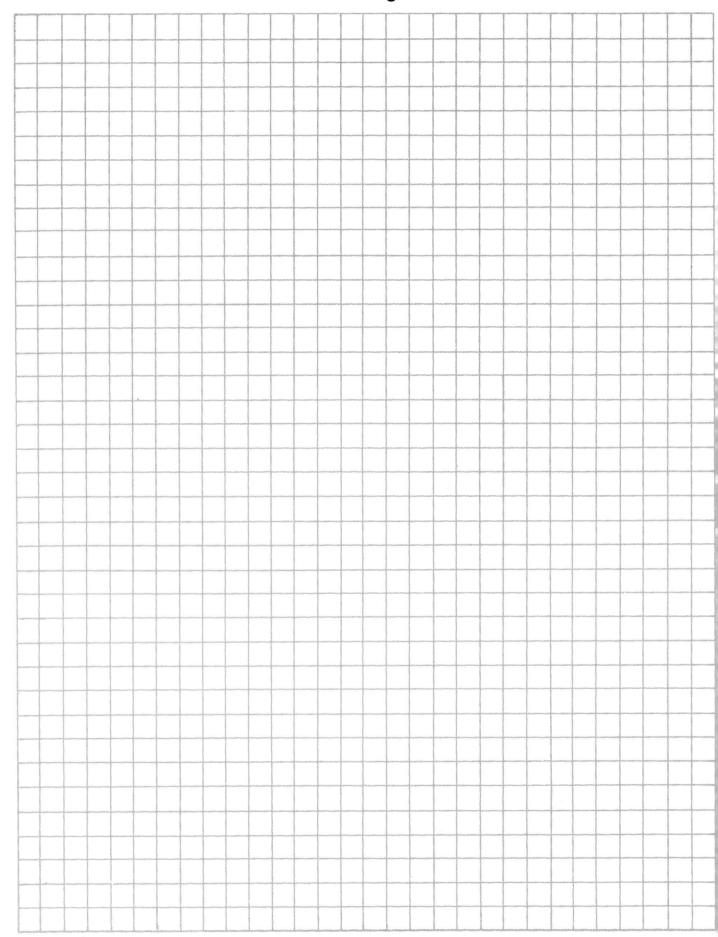

Since the column under consideration here has greater resistance to buckling than the case in Fig. 4-2e, the $(kL)_{cr}$ here will be greater than π. We can therefore evaluate Eq. (4-13) beginning with $kL = \pi$ and increasing it slowly until the value of the left side of Eq. (4-13) changes sign. Since $(kL)_{cr}$ lies between the values of kL for which the left side of Eq. (4-13) has opposite sign, we now have bounds on $(kL)_{cr}$. To obtain improved bounds, we take the average of the two bounding values, which we will designate by $(kL)_{av}$. If the value of the left side of Eq. (4-13) obtained by using $(kL)_{av}$ is positive (negative), then $(kL)_{cr}$ lies between $(kL)_{av}$ and that value of kL for which the left side of Eq. (4-13) is negative (positive). This process is continued, using the successive values of $(kL)_{av}$ to obtain improved bounds on $(kL)_{cr}$ until the desired accuracy is obtained.

The last two equations in Eq. (4-12) imply perfect rigidity of the surrounding structure with respect to rotation. A more general result may be obtained by taking into account the elasticity of the surrounding structure in this respect. Suppose that the equivalent torsional spring constants for the surrounding structure are $K(T, 0)$ and $K(T, L)$ at $x = 0$ and $x = L$, respectively. Then Eq. (4-12) is replaced by

$$y(L) = 0$$

$$M(0) = K(T, 0) \frac{dy(0)}{dx} \tag{4-15}$$

$$M(L) = -K(T, L) \frac{dy(L)}{dx}$$

Proceeding as before, with Eq. (4-15) replacing Eq. (4-12), we obtain the following equation for kL:

$$\left\{ \left[\frac{L^3}{EI(kL)^3} \right] [K(T, 0) + K(T, L)]K(0) - \left[\frac{L^4 K(0)K(T, 0)K(T, L)}{(EI)^2(kL)^3} \right] \right.$$

$$+ \left[\frac{K(0)L^2}{(kL)} \right] + \left[\frac{LK(T, 0)K(T, L)}{EI(kL)} \right] - \left[\frac{EI(kL)}{L} \right] \right\} \sin kL$$

$$+ \left\{ K(T, 0) + K(T, L) - \left[\frac{L^3}{EI(kL)^2} \right] [K(T, 0) + K(T, L)]K(0) \right\} \cos kL$$

$$+ 2 \left[\frac{L^4 K(0)K(T, 0) + K(T, L)}{(EI)^2(kL)^4} \right] (1 - \cos kL) = 0 \tag{4-16}$$

The lowest value of kL satisfying Eq. (4-16) is the $(kL)_{cr}$ to be substituted in Eq. (4-14) in order to obtain the critical load. Here there is no apparent good guess with which to begin computations. Considering the current accessibility of computers, a convenient approach would be to obtain a plot of the left side of Eq. (4-16) for $0 \leq kL < \pi$, and if there is no change in sign, extend the plot up to $kL = 2\pi$, which is the solution for the column with a perfectly rigid surrounding structure (Fig. 4-2c).

4-4 MODIFIED BUCKLING FORMULAS

The critical-load formulas developed above provide satisfactory values of the allowable load for very slender columns for which buckling, as manifested by unacceptably large deformation, will occur within the elastic range of the material. For more massive columns, the deformation enters the plastic region (where strain increases more rapidly with stress) prior to the onset of buckling. To take into account this change in the stress-strain relationship, we modify the Euler formula. We define the *tangent modulus E(t)* as the slope of the tangent to the stress-strain curve at a given strain. Then the modified formulas for the critical load are obtained by substituting $E(t)$ for E in Eq. (4-9) and Eq. (4-13) plus Eq. (4-14) or Eq. (4-16) plus Eq. (4-14). This will produce a more accurate prediction of the buckling load. However, this may not be the most desirable design approach. In general, a design which will produce plastic deformation under the operating load is undesirable. Hence, for a column which will undergo plastic deformation prior to buckling, the preferred design-limiting criterion is the onset of plastic deformation, not the buckling.

4-5 STRESS-LIMITING CRITERION

We will now develop a design criterion which will enable us to use the yield strength as the upper bound for acceptable design regardless of whether the stress at the onset of yielding precedes or follows buckling. Here we follow Ref. [4-1]. This approach has the advantage of providing a single bounding criterion that holds irrespective of the mode of failure. We begin by noting that, in general, real columns will have some imperfection, such as crookedness of the centroidal axis or eccentricity of the axial load. Figure 4-4 shows the difference between the behavior of an ideal, perfectly straight column subjected to an axial load, in which case we obtain a distinct critical point, and the behavior of a column with some imperfection.

It is clear from Fig. 4-4 that the load-deflection curve for an imperfect column has no distinct critical point. Instead, it has two distinct regions. For small axial loads, the deflection increases slowly with load. When the load is approaching the critical value obtained for a perfect column, a small increment in load produces a large change in deflection. These two regions are joined by a "knee." Thus the advent of buckling in a real column corresponds to the entry of the column into the second, above-the-knee, load-deflection region. A massive column will reach the stress at the yield point prior to buckling, so that the yield strength will be the limiting criterion for the maximum allowable load. A slender column will enter the above-the-knee region prior to reaching the stress at the yield point, but once in the above-the-knee region, it requires only a small increment in load to produce a sufficiently large increase in deflection to reach the yield point. Thus the corresponding yield load may

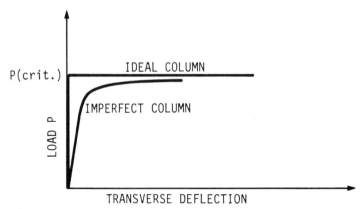

FIG. 4-4 Typical load-deflection curves for ideal and real columns.

be used as an adequate approximation of the buckling load for a slender column as well. Hence the yield strength provides an adequate design bound for both massive and slender columns. It is also important to note that, in general, columns found in applications are sufficiently massive that the linear theory developed here is valid within the range of deflection that is of interest.

Application of Eq. (4-1) to a simply supported imperfect column with constant properties over its length yields a modification of Eq. (4-4). Thus,

$$\frac{d^2y}{dx^2} + k^2y = k^2(e - Y) \tag{4-17}$$

where e = eccentricity of the axial load P (taken as positive in the positive y direction), and Y = initial deflection (crookedness) of the unloaded column. The x axis is taken through the end points of the centroidal axis, so that Eq. (4-5) still holds and Y is zero at the end points. Note that the functions in the right side of Eq. (4-8) form a basis for a trigonometric (Fourier) series, so that any function of interest may be expressed in terms of such a series. Thus we can write

$$Y = \sum_{n=1}^{\infty} c(n) \sin \frac{n\pi x}{L} \tag{4-18}$$

where
$$c(n) = \frac{2}{L} \int_0^L Y(x) \sin \frac{n\pi x}{L} \, dx \tag{4-19}$$

The solution for the deflection y in Eq. (4-17) is given by

$$y = P\sum_{n=1}^{\infty} \left[c(n) - \frac{4e}{n\pi} \right] \frac{\sin (n\pi x/L)}{[EI(n\pi/L)^2 - P]} \tag{4-20}$$

The maximum deflection $y_{\max}$ of a simply supported column will usually (except for cases with a pronounced and asymmetrical initial deformation or antisymmetrical load eccentricity) occur at the column midpoint. A good approximation (probably within 10 percent) of $y_{\max}$ in the above-the-knee region that may be used in deflection-limited column design is given by the coefficient of the first term in Eq. (4-20):

$$y_{\max} = \frac{P[c(1) - (4e/\pi)]}{EI(\pi/L)^2 - P} \tag{4-21}$$

The maximum bending moment will also usually occur at the column midpoint and at incipient yielding is closely approximated by

$$M_{\max} = P\left\{ e - Y_{\mathrm{mid}} + \left[\frac{4e}{\pi} - c(1) \right] \frac{P}{[EI(\pi/L)^2 - P]} \right\} \tag{4-22}$$

The immediately preceding analysis deals with the bending moment about the z axis (normal to the paper). Clearly, a similar analysis can be made with regard to bending about the y axis (Fig. 4-1). Unlike the analysis of the perfect column, where it is merely a matter of finding the buckling load about the weaker axis, in the present approach the effects about the two axes interact in a manner familiar from analysis of an eccentrically loaded short strut. We now use the familiar expression for combining direct axial stresses and bending stresses about two perpendicular axes. Since there is no ambiguity, we will suppress the negative sign associated with compressive stress:

$$\sigma = \frac{P}{A} + \frac{M(z)c(y)}{I(z)} + \frac{M(y)c(z)}{I(y)} \tag{4-23}$$

where $c(y)$ and $c(z)$ = perpendicular distances from the z axis and y axis, respectively (these axes meet the x axis at the cross-section centroid at the origin), to the outermost fiber in compression; A = cross-sectional area of the column; and σ = total compressive stress in the fiber which is farthest removed from both the y and z axes. For an elastic design limited by yield strength, σ is replaced by the yield strength; $M(z)$ in the right side of Eq. (4-23) is the magnitude of the right side of Eq. (4-22); and $M(y)$ is an expression similar to Eq. (4-22) in which the roles of the y and z axes interchange.

Usually, in elastic design, the yield strength is divided by a chosen factor of safety η to get a permissible or allowable stress. In problems in which the stress increases linearly with the load, dividing the yield stress by the factor of safety is equivalent to multiplying the load by the factor of safety. However, in the problem at hand, it is clear from the preceding development that the stress is not a linear function of the axial load and that we are interested in the behavior of the column as it enters the above-the-knee region in Fig. 4-4. Here it is necessary to multiply the applied axial load by the desired factor of safety. The same procedure applies in introducing a factor of safety in the critical-load formulas previously derived.

EXAMPLE 1. We will examine the design of a nominally straight column supporting a nominally concentric load. In such a case, a circular column cross section is the most reasonable choice, since there is no preferred direction. For this case, Eq. (4-23) reduces to

$$\sigma = \frac{P}{A} + \frac{Mc}{I} \tag{1}$$

For simplicity, we will suppose that the principal imperfection is due to the eccentric location of the load and that the column crookedness effect need not be taken into account, so that Eq. (4-22) reduces to

$$M_{\text{max}} = P\left\{ e + \left(\frac{4e}{\pi}\right) \frac{P}{[EI(\pi/L)^2 - P]} \right\} \tag{2}$$

Note that for a circular cross section of radius R, the area and moment of inertia are, respectively,

$$A = \pi R^2 \quad \text{and} \quad I = \frac{\pi R^4}{4} = \frac{A^2}{4\pi} \tag{3}$$

We will express the eccentricity of the load as a fraction of the cross-section radius. Thus,

$$e = \varepsilon R \tag{4}$$

Then we have, from Eqs. (1) through (4),

$$\sigma_{\text{allow}} = \frac{P}{A} + \frac{4P\varepsilon}{A}\left\{ 1 + \left(\frac{4}{\pi}\right) \frac{P}{[(EA^2)/(4\pi)](\pi/L)^2 - P]} \right\} \tag{5}$$

Usually P and L are given, σ and E are the properties of chosen material, and ε is determined from the clearances, tolerances, and kinematics involved, so that Eq. (5) is reduced to a cubic in A.

At the moment, however, we are interested in comparing the allowable nominal column stress P/A with the allowable stress of the material σ_{allow} for columns of different lengths. Keeping in mind that the radius of gyration r of a circular cross section of geometric radius R is $R/2$, we will define

$$\frac{R}{2} = r$$

$$\frac{\sigma_{\text{allow}}}{P/A} = p \qquad (6)$$

$$\frac{E}{\sigma_{\text{allow}}} = q$$

Then Eq. (5) may be written as

$$p = 1 + 4\varepsilon + \frac{16\varepsilon}{\pi[\pi^2 pq(r/L)^2 - 1]} \qquad (7)$$

The first term on the right side of Eq. (7) is due to direct compressive stress; the second term is due to the bending moment produced by the load eccentricity; the third term is due to the bending moment arising from the column deflection. When ε is small, p will be close to unity unless the denominator in the third term on the right side of Eq. (7) becomes small; that is, the moment due to the column deflection becomes large. The ratio L/r, whose reciprocal appears in the denominator of the third term, is called the *slenderness ratio*. Equation (7) may be rewritten as a quadratic in p. Thus,

$$\pi^2 q\left(\frac{r}{L}\right)^2 p^2 - \left[(1 + 4\varepsilon)\pi^2 q\left(\frac{r}{L}\right)^2 + 1\right]p + (1 + 4\varepsilon) - \frac{16\varepsilon}{\pi} = 0 \qquad (8)$$

We will take for q the representative value of 1000 and tabulate $1/p$ for a number of values of L/r and ε. To compare the value of $1/p$ obtained from Eq. (8) with the corresponding result from Euler's formula, we will designate the corresponding result obtained by Euler's formula as $1/p_{\text{cr}}$ and recast Eq. (4-9) as

$$\frac{1}{p_{\text{cr}}} = \pi^2 q\left(\frac{r}{L}\right)^2 \qquad (9)$$

To interpret the results in Table 4-1, note that the quantities in the second and third columns of the table are proportional to the allowable loads calculated from the respective equations. As expected, the Euler formula is completely inapplicable when L/r is 50. Also, as expected, the allowable load decreases as the eccentricity increases. However, the effect of eccentricity on the allowable load decreases as the slenderness ratio L/r increases. Hence when L/r is 250, the Euler buckling load, which is the limiting case for which the eccentricity is zero, is only about 2 percent higher than when the eccentricity is 2 percent.

Notes ▪ Drawings ▪ Ideas

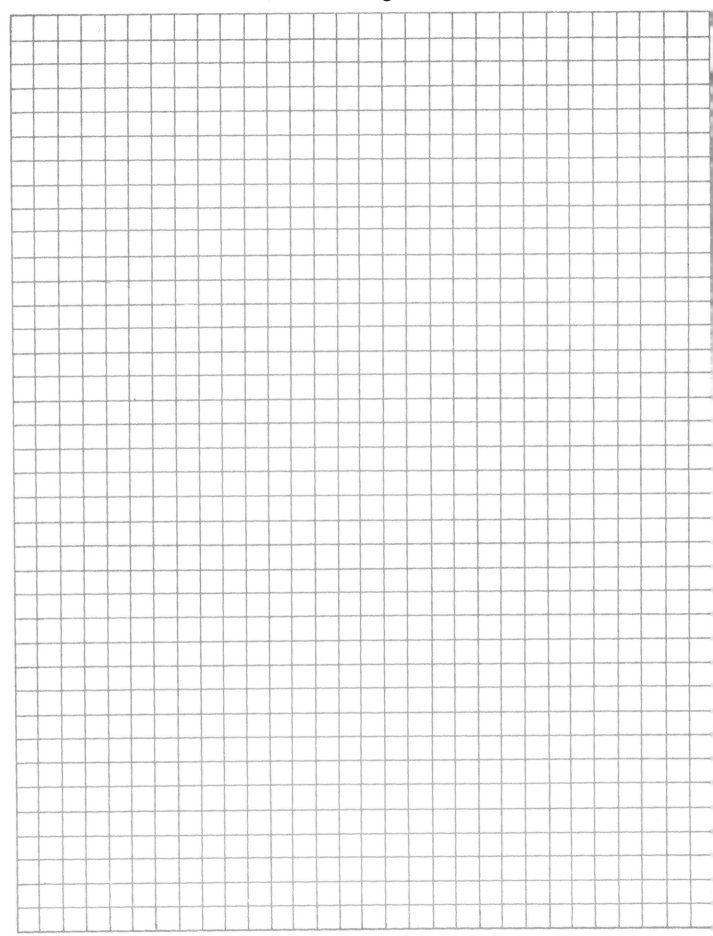

TABLE 4-1 Influence of Eccentricity and
Slenderness Ratio on Allowable Load

ε	L/r	p^{-1}	p_{cr}^{-1}
0.02	50	0.901	3.95
	150	0.408	0.439
	250	0.155	0.158
0.05	50	0.791	3.95
	150	0.374	0.439
	250	0.151	0.158
0.10	50	0.665	3.95
	150	0.333	0.439
	250	0.145	0.158

4-6 BEAM-COLUMN ANALYSIS

A member that is subjected to both a transverse load and an axial load is frequently called a *beam-column*. To apply the immediately preceding stress-limiting criterion to a beam-column, we first determine the moment distribution, say, M_{tr}, and the corresponding deflection, say, Y_{tr}, resulting from the transverse load acting alone. Suppose that the transverse load is symmetrical about the column midpoint, and let $Y_{tr, mid}$ and $M_{tr, mid}$ be the values of Y_{tr} and M_{tr} at the column midpoint. Then the only modifications necessary in the preceding development are to replace Y by $Y + Y_{tr}$ in Eqs. (4-17) and (4-20), as well as to replace Y_{mid} by $Y_{mid} + Y_{tr, mid}$ and add $M_{tr, mid}$ on the right side of Eq. (4-23). If the transverse load is not symmetrical, then it is necessary to determine the maximum moment by using an approach which will now be developed.

Note that, at any point, x, the moment about the z axis is

$$M(z) = P(e - y - Y) + M(z)_{tr} \qquad (4\text{-}24)$$

where Y includes the deflection due to $M(z)_{tr}$. $M(y)$ has the same form as Eq. (4-24) but with the roles of y and z interchanged. The maximum stress for any given value of x is given by Eq. (4-23). We seek to apply this equation at that value of x which yields the maximum value of σ. A method that is reasonably efficient in locating a minimum or maximum to any desired accuracy is the golden-section search. However, this method is limited to finding the minimum (maximum) of a unimodal function, that is, a function which has only one minimum (maximum) in the interval in which the search is conducted. We therefore have to conduct some exploratory calculation to find the stress at, say, a dozen points on the beam-column in order to locate the unimodal interval of interest within which to apply the golden-section search. The actual number of exploratory calculations will depend on the individual case. For example, in a simply supported case with a unimodal transverse moment, there is clearly only one maximum. But, in general, we must check enough points to be sure that a potential maximum is not overlooked.

The golden-section search procedure is as follows: Suppose that we seek the minimum value of $F(x)$ in Fig. 4-5 within the interval D (note that if we sought a maximum in Fig. 4-5, we would have to conduct two searches). We locate two points $x(1)$ and $x(2)$. The first is $0.382D$ from the left end of the interval; the second is

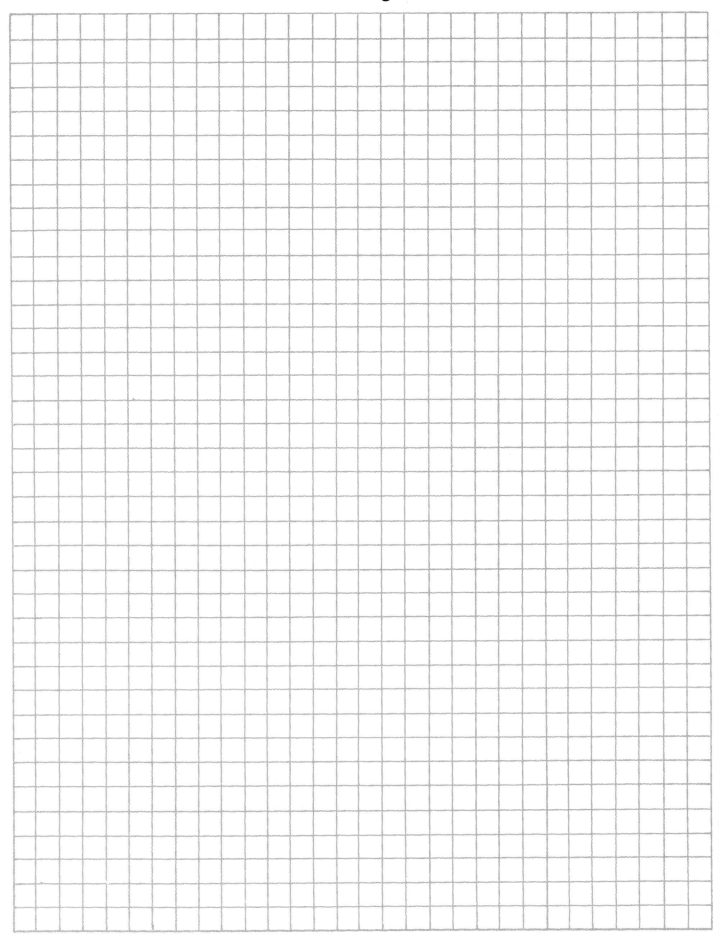

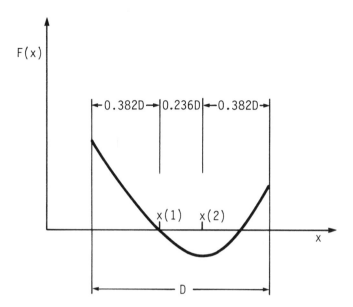

FIG. 4-5   Finding min $F(x)$ in the interval D.

0.382D from the right end. We evaluate $F(x)$ at the chosen points. If the value of $F(x)$ at $x(1)$ is algebraically smaller than at $x(2)$, then we eliminate the subinterval between $x(2)$ and the right end of the search interval. In the oppositive event, we eliminate the subinterval between $x(1)$ and the left end. The distance of 0.236D between $x(1)$ and $x(2)$ equals 0.382 of the new search interval, which is 0.618D overall. Thus we have already located one point at 0.382 of the new search interval, and we have determined the value of $F(x)$ at that point. We need only locate a point that is 0.382 from the opposite end of the new search interval, evaluate $F(x)$ at that point, and go through the elimination procedure, as before, to further reduce the search interval. The process is repeated until the entire remaining interval is small and the value of $F(x)$ at the two points at which it is evaluated is the same within the desired accuracy. The preceding method may run into difficulty if the function $F(x)$ is flat, that is, if $F(x)$ does not vary over a substantial part of the search interval or if $F(x)$ happens to have the same value at the two points that are compared with each other. The first case is not likely to occur in the types of problems under consideration here. In the second case, we calculate $F(x)$ at another point, near one of the two points of comparison, and use the newly chosen point in place of the nearby original point. Reference [4-2] derives the golden section search strategy and provides equations (p. 289) for predicting the number of function evaluations required to attain a specified fractional reduction of search interval, or an absolute final search interval size.

4-7 APPROXIMATE METHOD

The reason why we devoted so much attention to uniform prismatic column problems is that their solution is analytically simple, so that we could obtain the results directly. In most other cases, we have to be satisfied with approximate formulations for computer (or programmable calculator) calculation of the solution. The finite-element method is the approximation method most widely used in engineering at the present time to reduce problems dealing with continuous systems, such as beams and columns, to sets of algebraic equations that can be solved on a digital computer. When there is a single independent variable involved (as in our case), the interval of interest of the independent variable is divided into a set of subintervals called *finite elements*. Within each subinterval, the solution is represented by an arc that is defined by a simple function, usually a polynomial of low degree. The curve resulting from the connected arc segments should have a certain degree of smoothness (for the problems under discussion, the deflection curve and its first derivative should be continuous) and should approximate the solution. An effective method of obtaining a good approximation to the solution is based on the mechanical energy involved in the deformation process.

4-8 INSTABILITY OF BEAMS

Beams that have rectangular cross sections with the thickness much smaller than the depth are prone to instability involving rotation of the beam cross section about the beam axis. This tendency to instability arises because such beams have low resistance to torsion about their axes. In preparation for the analysis of this problem, recall that for a circular cylindrical member of length L and radius R, subjected to an axial torque T, the angle of twist ϕ is given by

$$\phi = \frac{TL}{JG} \qquad (4\text{-}25)$$

where G = shear modulus, and J = polar moment of inertia of the cross section. For noncircular cross sections, the form of the right side of Eq. (4-25) does not change; the only change is in the expression for J, the torsion constant of the cross section. If the thickness of the cross section, to be denoted by t, is small and does not vary much, then J is given by

$$3J = \int_0^l t^3 \, ds \qquad (4\text{-}26)$$

where s = running coordinate measured from one end of the cross section, and l = total developed length of the cross section. Thus for a rectangular cross section of depth h,

$$J = \frac{ht^3}{3} \qquad (4\text{-}27)$$

Clearly, J decreases rapidly as t decreases. To study the effect of this circumstance on beam stability, we will examine the deformation of a beam with rectangular cross section subjected to end moments $M(0)$, taking into account rotation of the cross section about the beam axis. The angle of rotation ϕ is assumed to be small, so that $\sin \phi$ may be replaced by ϕ and $\cos \phi$ by unity.

The equations of static equilibrium may be written from Fig. 4-6, where the moments are shown as vectors, using the right-hand rule:

$$-M'(z) + M(0) = 0$$
$$M'(y) + M(0)\phi = 0 \qquad (4\text{-}28)$$
$$\frac{dT}{dx} + \frac{dM'(z)}{dx} = 0$$

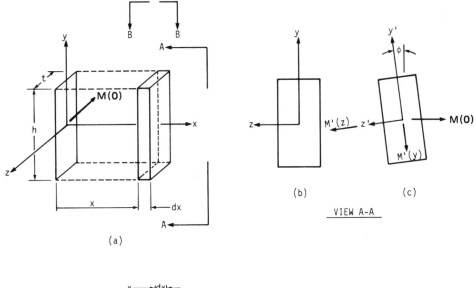

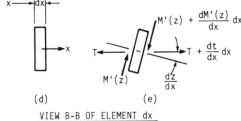

FIG. 4-6 Instability of beams. (*a*) Segment of a beam with applied moment $M(0)$ at the ends; (*b*) cross section of undeformed beam; (*c*) cross section after onset of beam instability; (*d*) top view of undeformed differential beam element; (*e*) the element after onset of instability.

These, combined with Eq. (4-25) and the standard moment-curvature relation as given by Eq. (4-1), lead to the defining differential equation (4-29) for angle ϕ:

$$GJ \frac{d^2\phi}{dx^2} + M^2(0)\phi \left[\frac{1}{EI(y)} - \frac{1}{EI(z)} \right] = 0 \qquad (4\text{-}29)$$

Suppose that the end faces of the beam are fixed against rotation about the x axis. Then the boundary conditions are

$$\phi(0) = \phi(L) = 0 \qquad (4\text{-}30)$$

Noting the similarity between Eq. (4-4) with its boundary conditions [Eq. (4-5)] and Eq. (4-29) with its boundary conditions [Eq. (4-30)], the similarity of the solutions is clear. Thus we obtain the expression for $M(0)_{\text{cr}}$.

$$[M(0)_{\text{cr}}]^2 = \frac{\pi^2 E}{L^2} GJ \left[\frac{I(z)I(y)}{I(z) - I(y)} \right] \qquad (4\text{-}31)$$

It may be seen from Eq. (4-31) that if the torsion constant J is small, the critical moment is small. In addition, it may be seen from the bracketed term in Eq. (4-31) that as the cross section approaches a square shape, the denominator becomes small, so that the critical moment becomes very large. The disadvantage of a square cross section is, of course, well known. To demonstrate it explicitly, we rewrite the expression for stress in a beam with rectangular cross section t by h:

$$\sigma = \frac{M(h/2)}{th^3/12} \tag{4-32}$$

in the form

$$th = \frac{6M}{\sigma h} \tag{4-33}$$

Thus, for given values of M and σ, the cross-sectional area required decreases as we increase h. Hence the role of Eq. (4-31) is to define the constraint on the maximum allowable depth-to-thickness ratio. The situation is similar for flanged beams. Here we have obtained the results for a simple problem to illustrate the disadvantage involved and the caution necessary in designing beams with thin-walled open cross sections. Implicit in this is the advantage of using, when possible, closed cross sections, such as box beams, which have a high torsional stiffness.

As we have seen, when the applied moment is constant over the entire length of the beam, the problem of definition and its solution have the same form as for the column-buckling problem. We can also have similiar types of boundary conditions. The boundary conditions used in Eq. (4-30) correspond to a simply supported column. If dy/dx and dz/dx are equal to zero at $x = 0$ for all y and z, then $d\phi/dx$ is equal to zero at $x = 0$. If this condition is combined with $\phi(0) = 0$, then we have the equivalent of a clamped column end. Hence we can use here the concept of equivalent beam length in the same manner as we used the equivalent column length before. In case a beam is subjected to transverse loads, so that the applied moment varies with x, the problem is more complex. For the proportioning of flanged beams, Ref. [4-3] should be used as a guide. This reference deals with structural applications, so that the size range of interest dealt with is different from the size range of interest in machine design. But the underlying principles of beam stability are the same, and the proportioning of the members should be similar.

EXAMPLE 2. We will examine the design of a beam of length L and rectangular cross section t by h. The beam is subjected to an applied moment M (we will not use any modifying symbols here, since there is no ambiguity), which is constant over the length of the beam. As noted previously, the required cross-sectional area th will decrease as h is increased. We take the allowable stress in the material to be σ. The calculated stress in the beam is not to exceed this value. Thus,

$$\sigma \geq \frac{M(h/2)}{th^3/12} \qquad \sigma \geq \frac{6M}{th^2} \qquad th \geq \frac{6M}{h\sigma} \tag{1}$$

We want the cross-sectional area th as small as possible. Hence h should be as large as possible. We can, therefore, replace the inequality in Eq. (1) by the equality

$$th = \frac{6M}{h\sigma} \tag{2}$$

The maximum value of h that we can use is subject to a constraint based on Eq. (4-31). We will use a factor of safety η in this connection. Thus,

$$(\eta M)^2 \leq \frac{\pi^2 E}{L^2} GJ\left[\frac{I(z)I(y)}{I(z) - I(y)}\right] \tag{3}$$

Here

$$I(z) = \frac{th^3}{12} \qquad I(y) = \frac{ht^3}{12} \qquad J = \frac{ht^3}{3} \tag{4}$$

Using Eq. (4), we may write Eq. (3) as

$$(\eta M)^2 \leq \frac{\pi^2 EG}{(6L)^2}\left(\frac{t}{h}\right)^2\left[\frac{1}{1 - (t/h)^2}\right](th)^4 \tag{5}$$

or, from Eq. (2),

$$(\eta M)^2 \leq \frac{\pi^2 EG}{(6L)^2}\left(\frac{6M}{h^3\sigma}\right)^2\left[\frac{1}{1 - [(6M/(h^3\sigma)]^2}\right]\left(\frac{6M}{h\sigma}\right)^4 \tag{6}$$

Since we seek to minimize th and maximize h, it may be seen from Eqs. (5) and (6) that the inequality sign may be replaced by the equality sign in those two equations. In Eq. (6), h is the only unspecified quantity. Further, since the square of t/h may be expected to be small compared to unity, we can obtain substantially simpler approximations of reasonable accuracy. As a first step, we have

$$\frac{1}{1 - (t/h)^2} = 1 + \left(\frac{t}{h}\right)^2 + \left(\frac{t}{h}\right)^4 + \cdots \tag{7}$$

If we only retain the first two terms in the right side of Eq. (7), we have

$$\eta M = \frac{\pi(EG)^{1/2}}{6L}\left[1 + \left(\frac{6M}{h^3\sigma}\right)^2\right]\left(\frac{6M}{h^3\sigma}\right)\left(\frac{6M}{h\sigma}\right)^2 \tag{8}$$

If we also neglect the square of t/h in comparison with unity, we obtain

$$h = \left[\frac{\pi(6M)^2(EG)^{1/2}}{\eta L\sigma^3}\right]^{1/5} \tag{9}$$

as a reasonable first approximation. Thus if we take the factor of safety η as 1.5, we have, for a steel member with $E = 30$ Mpsi, $G = 12$ Mpsi, and $\sigma = 30$ kpsi,

$$h = \left[\frac{\pi(36)[(30 \times 10^6)(12 \times 10^6)]^{1/2}}{(1.5)(30\,000)}\frac{M^2}{L}\right]^{1/5} = 8.62\left(\frac{M^2}{L}\right)^{1/5} \tag{10}$$

This is a reasonable approximation to the optimal height of the beam cross section. It may also be used as a starting point for an iterative solution to the exact expression, Eq. (6). For the purpose of iteration, we rewrite Eq. (6) as

$$h = \left\{\frac{\pi(6M)^2}{\eta L\sigma^3}\left[\frac{EG}{1 - [(6M)/(h^3\sigma)]^2}\right]^{1/2}\right\}^{1/5} \tag{11}$$

The value of h obtained from Eq. (9) is substituted into the right side of Eq. (11). The resultant value of h thus obtained is then resubstituted into the right side of Eq.

Notes · Drawings · Ideas

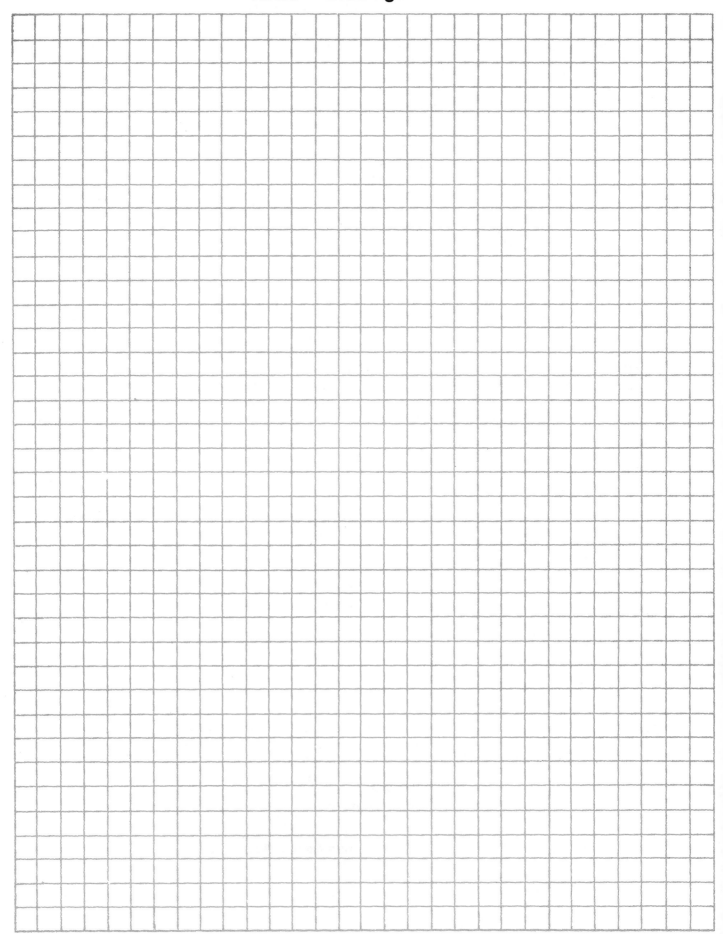

(11); the iterative process is continued until the computed value of h coincides with the value substituted into the right side to the desired degree of accuracy. Having determined h, we can determine t from Eq. (2).

REFERENCES

4-1 Herman, H., "On the Analysis of Uniform Prismatic Columns," *Transactions of the ASME, Journal of Mechanical Design,* vol. 103, 1981, pp. 274–276.

4-2 Mischke, C. R., *Mathematical Model Building,* 2d rev. ed., Iowa State University Press, Ames, Iowa, 1980.

4-3 *Manual of Steel Construction,* 8th ed., American Institute of Steel, Construction, New York.

Notes ▪ Drawings ▪ Ideas

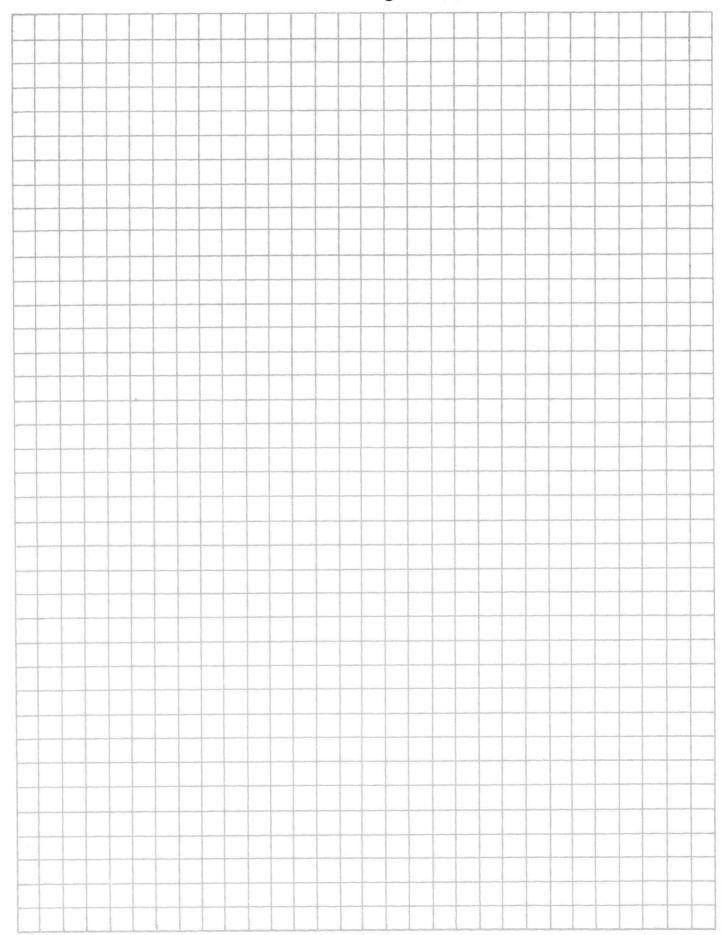

chapter **5**
CURVED BEAMS AND RINGS

JOSEPH E. SHIGLEY

Professor Emeritus
The University of Michigan
Ann Arbor, Michigan

NOTATION

A Area, or a constant

B Constant

C Constant

E Modulus of elasticity

e Eccentricity

F Force

G Modulus of rigidity

I Second moment of area (Table 1-1)

K Shape constant (Table 2-2), or second polar moment of area

M Bending moment

P Reduced load

Q Fictitious force

R Force reaction

r Ring radius

$\bar{r}$ Centroidal ring radius

T Torsional moment

U Strain energy

V Shear force

W Resultant of a distributed load

w Unit distributed load

X Constant

Y Constant

y Deflection

Z Constant

γ Load angle

ϕ Span angle, or slope

σ Normal stress

θ Angular coordinate or displacement

Methods of computing the stresses in curved beams for a variety of cross sections are included in this chapter. Rings and ring segments loaded normal to the plane of the ring are analyzed for a variety of loads and span angles, and formulas given for bending moment, torsional moment, and deflection.

5-1 BENDING IN THE PLANE OF CURVATURE

The distribution of stress in a curved member subjected to a bending moment in the plane of curvature is hyperbolic ([5-1], [5-2]) and is given by the equation

$$\sigma = \frac{My}{Ae(r - e - y)} \tag{5-1}$$

where r = radius to centroidal axis
y = distance from neutral axis
e = shift in neutral axis due to curvature (as noted in Table 5-1)

The moment M is computed about the *centroidal axis,* not the neutral axis. The maximum stresses, which occur on the extreme fibers, may be computed using the formulas of Table 5-1.

In most cases, the bending moment is due to forces acting to one side of the section. In such cases, be sure to add the resulting axial stress to the maximum stresses obtained using Table 5-1.

5-2 CASTIGLIANO'S THEOREM

A complex structure loaded by any combination of forces, moments, and torques can be analyzed for deflections by using the elastic energy stored in the various components of the structure [5-1]. The method consists of finding the total strain energy stored in the system by all the various loads. Then the displacement corresponding to a particular force is obtained by taking the partial derivative of the total energy with respect to that force. This procedure is called *Castigliano's theorem.* General expressions may be written as

$$y_i = \frac{\partial U}{\partial F_i} \qquad \theta_i = \frac{\partial U}{\partial T_i} \qquad \phi_i = \frac{\partial U}{\partial M_i} \tag{5-2}$$

where U = strain energy stored in structure
y_i = displacement of point of application of force F_i in the direction of F_i
θ_i = angular displacement at T_i
ϕ_i = slope or angular displacement at moment M_i

If a displacement is desired at a point on the structure where no force or moment exists, then a fictitious force or moment is placed there. When the expression for the corresponding displacement is developed, the fictitious force or moment is equated to zero, and the remaining terms give the deflection at the point where the fictitious load had been placed.

Castigliano's method can also be used to find the reactions in indeterminate structures. The procedure is simply to substitute the unknown reaction in Eq. (5-2) and use zero for the corresponding deflection. The resulting expression then yields the value of the unknown reaction.

It is important to remember that the displacement-force relation must be linear. Otherwise, the theorem is not valid.

TABLE 5-1 Eccentricities and Stress Factors for Curved Beams†

1. Rectangle

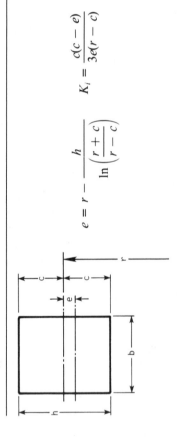

$$e = r - \frac{h}{\ln\left(\dfrac{r+c}{r-c}\right)} \qquad K_i = \frac{c(c-e)}{3e(r-c)} \qquad K_o = \frac{c(c+e)}{3e(r+c)}$$

2. Solid round

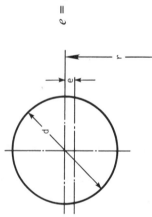

$$e = r - \frac{d^2}{4(2r - \sqrt{4r^2 - d^2})} \qquad K_i = \frac{d(d-2e)}{8e(2r-d)} \qquad K_o = \frac{d(d+2e)}{8e(2r+d)}$$

3. Hollow round

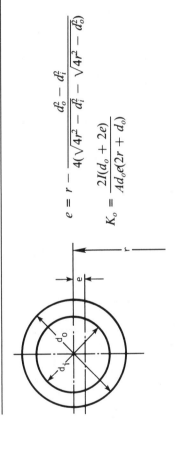

$$e = r - \frac{d_o^2 - d_i^2}{4(\sqrt{4r^2 - d_i^2} - \sqrt{4r^2 - d_o^2})}$$

$$K_o = \frac{2I(d_o + 2e)}{Ad_o e(2r + d_o)} \qquad K_i = \frac{2I(d_o - 2e)}{Ad_o e(2r - d_o)}$$

4. Hollow rectangle

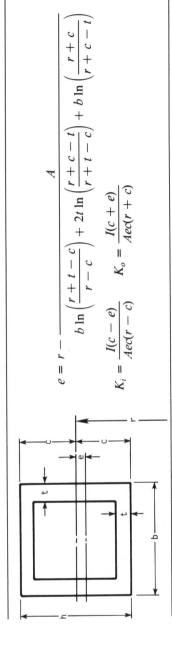

$$e = r - \frac{A}{b\ln\left(\dfrac{r + t - c}{r - c}\right) + 2t\ln\left(\dfrac{r + c - t}{r + t - c}\right) + b\ln\left(\dfrac{r + c}{r + c - t}\right)}$$

$$K_i = \frac{I(c - e)}{Aec(r - c)} \qquad K_o = \frac{I(c + e)}{Aec(r + c)}$$

TABLE 5-1 Eccentricities and Stress Factors for Curved Beams† (*Continued*)

5. Trapezoid

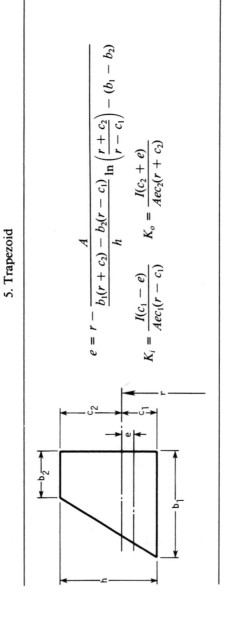

$$e = r - \cfrac{A}{\cfrac{b_1(r + c_2) - b_2(r - c_1)}{h} \ln\left(\cfrac{r + c_2}{r - c_1}\right) - (b_1 - b_2)}$$

$$K_i = \frac{I(c_1 - e)}{Aec_1(r - c_1)} \qquad K_o = \frac{I(c_2 + e)}{Aec_2(r + c_2)}$$

6. T Section

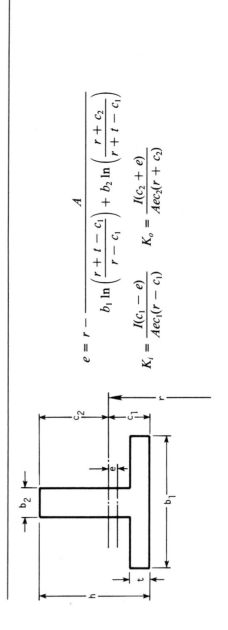

$$e = r - \cfrac{A}{b_1 \ln\left(\cfrac{r + t - c_1}{r - c_1}\right) + b_2 \ln\left(\cfrac{r + c_2}{r + t - c_1}\right)}$$

$$K_i = \frac{I(c_1 - e)}{Aec_1(r - c_1)} \qquad K_o = \frac{I(c_2 + e)}{Aec_2(r + c_2)}$$

7. U Section

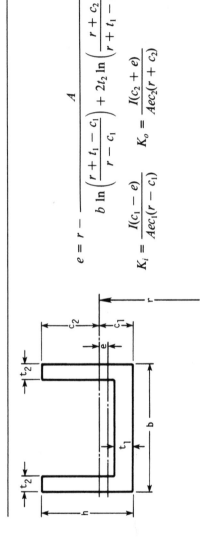

$$e = r - \cfrac{A}{b \ln\left(\cfrac{r + t_1 - c_1}{r - c_1}\right) + 2t_2 \ln\left(\cfrac{r + c_2}{r + t_1 - c_1}\right)}$$

$$K_i = \frac{I(c_1 - e)}{Aec_1(r - c_1)} \qquad K_o = \frac{I(c_2 + e)}{Aec_2(r + c_2)}$$

†Notation: r = radius of curvature to centroidal axis of section; A = area; I = second moment of area; e = distance from centroidal axis to neutral axis; $\sigma_i = K_i\sigma$ and $\sigma_o = K_o\sigma$ where σ_i and σ_o are the normal stresses on the fibers having the smallest and largest radii of curvature, respectively, and σ are the corresponding stresses computed on the same fibers of a straight beam. (Formulas for A and I can be found in Table 1-1.)

Notes ▪ Drawings ▪ Ideas

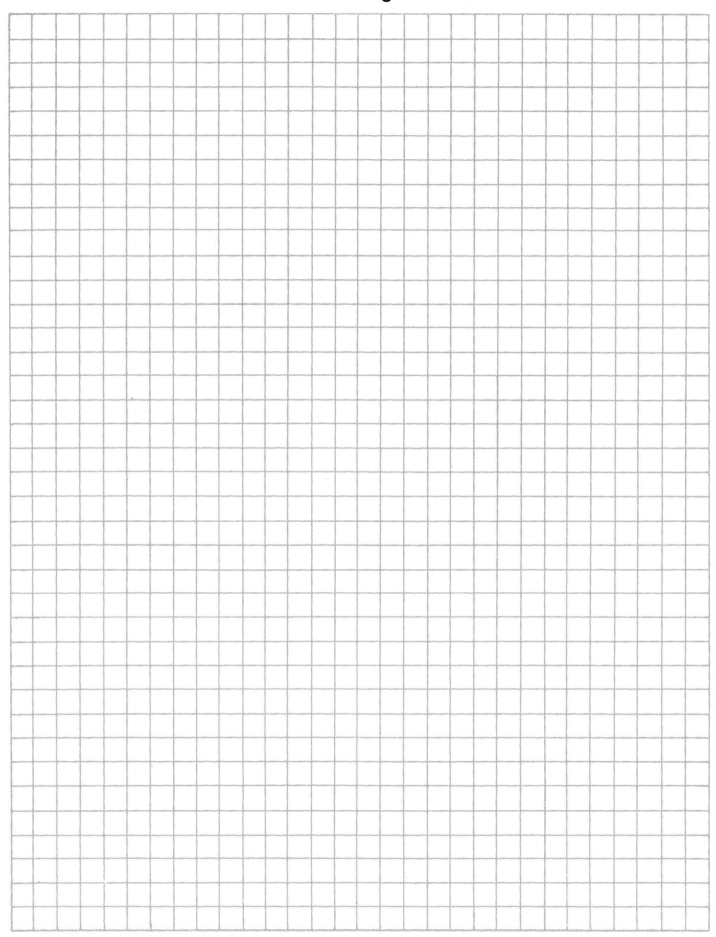

TABLE 5-2 Strain Energy Formulas

Loading	Formula
1. Axial force F	$U = \dfrac{F^2 l}{2AE}$
2. Shear force F	$U = \dfrac{F^2 l}{2AG}$
3. Bending moment M	$U = \displaystyle\int \dfrac{M^2 \, dx}{2EI}$
4. Torsional moment T	$U = \dfrac{T^2 l}{2GK}$

5-3 RING SEGMENTS WITH ONE SUPPORT

Figure 5-1 shows a cantilevered ring segment fixed at C. The force F causes bending, torsion, and direct shear. The moments and torques at the fixed end C and at any

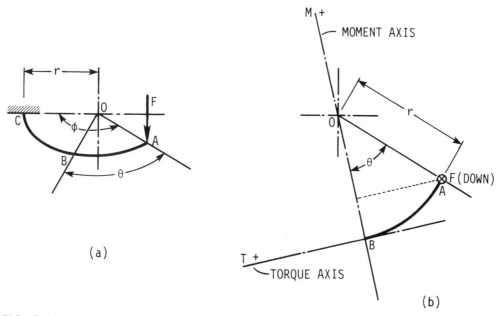

(a)

(b)

FIG. 5-1 (*a*) Ring segment of span angle ϕ loaded by force F normal to the plane of the ring. (*b*) View of portion of ring AB showing positive directions of the moment and torque for section at B.

TABLE 5-3 Formulas for Ring Segments with One Support

Loading	Term	Formula
End load F	Moment Torque	$M = Fr \sin \theta \qquad M_C = Fr \sin \phi$ $T = Fr(1 - \cos \theta) \qquad T_C = Fr(1 - \cos \phi)$
	Derivatives	$\dfrac{\partial M}{\partial F} = r \sin \theta \qquad \dfrac{\partial T}{\partial F} = r(1 - \cos \theta)$
	Deflection coefficients	$A = \phi - \sin \phi \cos \phi$ $B = 3\phi - 4 \sin \phi + \sin \phi \cos \phi$
Distributed load w; fictitious load Q	Moment Torque	$M = wr^2(1 - \cos \theta) \qquad M_C = wr^2(1 - \cos \phi)$ $T = wr^2(\theta - \sin \theta) \qquad T_C = wr^2(\phi - \sin \phi)$
	Derivatives	$\dfrac{\partial M}{\partial Q} = r \sin \theta \qquad \dfrac{\partial T}{\partial Q} = r(1 - \cos \theta)$
	Deflection coefficients	$A = 2 - 2 \cos \phi - \sin^2 \phi$ $B = \phi^2 - 2\phi \sin \phi + \sin^2 \phi$

section B are shown in Table 5-3. The shear at C is $R_C = F$. Stresses in the ring can be computed using the formulas of Chap. 2.

To obtain the deflection at end A we use Castigliano's theorem. Neglecting direct shear and noting from Fig. 5-1b that $l = r \, d\theta$, we determine the strain energy from Table 5-2 to be

$$U = \int_0^\phi \frac{M^2 r \, d\theta}{2EI} + \int_0^\phi \frac{T^2 r \, d\theta}{2GK} \qquad (5\text{-}3)$$

Then the deflection y at A and in the direction of F is computed from

$$y = \frac{\partial U}{\theta F} = \frac{r}{EI} \int_0^\phi M \frac{\partial M}{\partial F} \, d\theta + \frac{r}{GK} \int_0^\phi T \frac{\partial T}{\partial F} \, d\theta \qquad (5\text{-}4)$$

The terms for this relation are shown in Table 5-3. It is convenient to arrange the solution in the form

$$y = \frac{Fr^3}{2} \left(\frac{A}{EI} + \frac{B}{GK} \right) \qquad (5\text{-}5)$$

where the coefficients A and B are related only to the span angle. These are listed in Table 5-3.

Figure 5-2a shows another cantilevered ring segment loaded now by a distributed load. The resultant load is $W = wr\phi$; a shear reaction $R = W$ acts upward at the fixed end C, in addition to the moment and torque reactions shown in Table 5-3.

A force $W = wr\theta$ acts at the centroid of segment AB in Fig. 5-2b. The centroidal radius is

$$\bar{r} = \frac{2r \sin (\theta/2)}{\theta} \qquad (5\text{-}6)$$

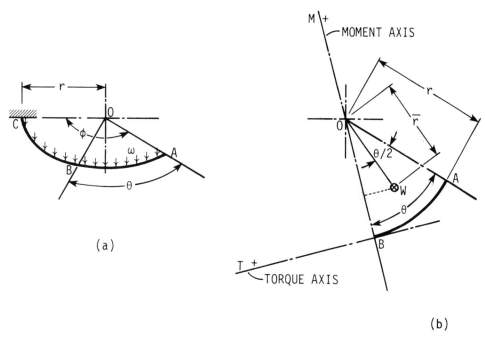

FIG. 5-2 (*a*) Ring segment of span angle ϕ loaded by a uniformly distributed load *w* acting normal to the plane of the ring segment; (*b*) view of portion of ring *AB*; force *W* is the resultant of the distributed load *w* acting on portion *AB* of ring, and it acts at the centroid.

To determine the deflection of end *A* we employ a fictitious force *Q* acting down at end *A*. Then the deflection is

$$y = \frac{\partial U}{\partial Q} = \frac{r}{EI} \int_0^{\phi} M \frac{\partial M}{\partial Q} \, d\theta + \frac{r}{GK} \int_0^{\phi} T \frac{\partial T}{\partial Q} \, d\theta \qquad (5\text{-}7)$$

The components of the moment and torque due to *Q* can be obtained by substituting *Q* for *F* in the moment and torque equations in Table 5-3 for an end load *F*; then the total of the moments and torques is obtained by adding this result to the equations for *M* and *T* due only to the distributed load. When the terms in Eq. (5-7) have been formed, the force *Q* can be placed equal to zero prior to integration. The deflection equation can then be expressed as

$$y = \frac{wr^4}{2} \left(\frac{A}{EI} + \frac{B}{GK} \right) \qquad (5\text{-}8)$$

5-4 RINGS WITH SIMPLE SUPPORTS

Consider a ring loaded by any set of forces *F* and supported by reactions *R*, all normal to the ring plane, such that the force system is statically determinate. The system shown in Fig. 5-3, consisting of five forces and three reactions, is statically determinate and is such a system. By choosing an origin at any point *A* on the ring, all

Notes ▪ Drawings ▪ Ideas

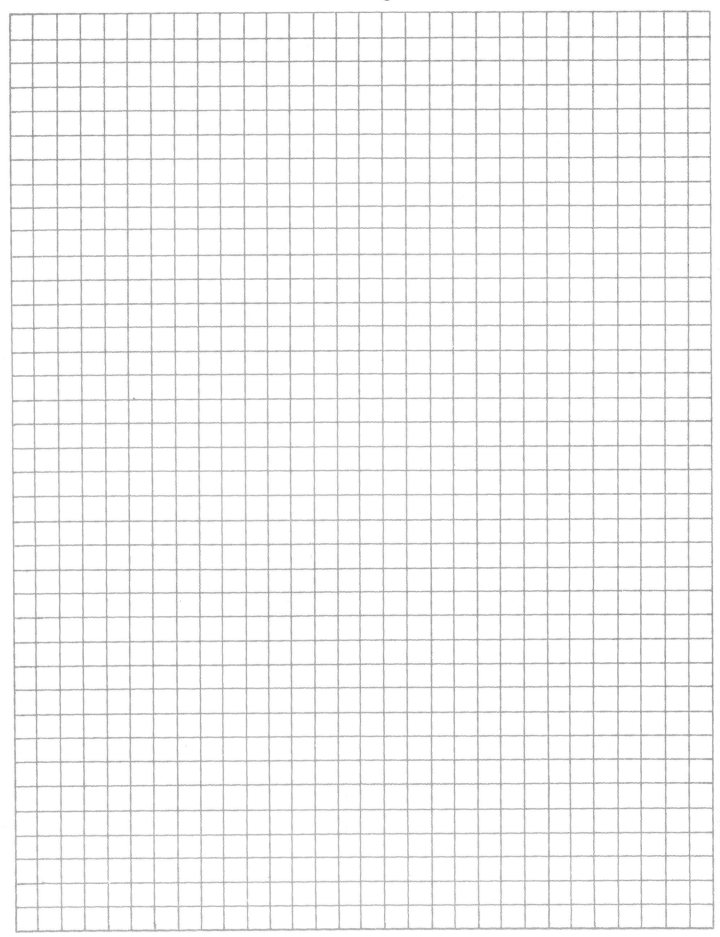

forces and reactions can be located by the angles ϕ measured counterclockwise from A. By treating the reactions as negative forces, Den Hartog [5-3], pp. 319–323, describes a simple method of determining the shear force, the bending moment, and the torsional moment at any point on the ring. The method is called *Biezeno's theorem*.

A term called the *reduced load P* is defined for this method. The reduced load is obtained by multiplying the actual load, plus or minus, by the fraction of the circle corresponding to its location from A. Thus for a force F_i the reduced load is

$$P_i = \frac{\phi_i}{360°} F_i \qquad (5\text{-}9)$$

Then Biezeno's theorem states that the shear force V_A, the moment M_A, and the torque T_A at section A, all statically indeterminate, are found from the set of equations

$$V_A = \sum_n P_i$$

$$M_A = \sum_n P_i r \sin \phi_i \qquad (5\text{-}10)$$

$$T_A = \sum_n P_i r (1 - \cos \phi_i)$$

where n = number of forces and reactions together. The proof uses Castigliano's theorem and may be found in Ref. [5-3].

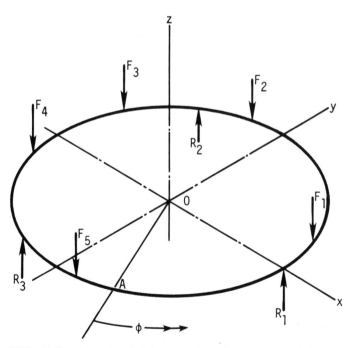

FIG. 5-3 Ring loaded by a series of concentrated forces.

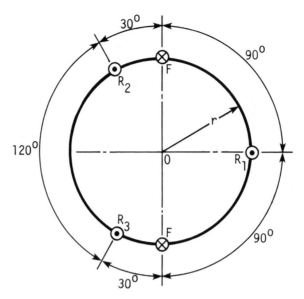

FIG. 5-4   Ring loaded by the two forces F and supported by reactions R_1, R_2, and R_3. The crosses indicate that the forces act downward; the heavy dots at the reactions R indicate an upward direction.

EXAMPLE 1. Find the shear force, bending moment, and torsional moment at the location of R_3 for the ring shown in Fig. 5-4.

Solution. Using the principles of statics, we first find the reactions to be

$$R_1 = R_2 = R_3 = \tfrac{2}{3}F$$

Choosing point A at R_3, the reduced loads are

$$P_0 = -\frac{0°}{360°} R_3 = 0 \qquad P_1 = \frac{30}{360} F = 0.0833F$$

$$P_2 = -\frac{120}{360} R_1 = -\frac{120}{360}\frac{2}{3} F = -0.2222F$$

$$P_3 = \frac{210}{360} F = 0.5833F$$

$$P_4 = -\frac{240}{360} R_2 = -\frac{240}{360}\frac{2}{3} F = -0.4444F$$

Then, using Eq. (5-10), we find $V_A = 0$. Next,

$$M_A = \sum_5 P_i r \sin \phi_i$$

$$= Fr (0 + 0.0833 \sin 30° - 0.2222 \sin 120° + 0.5833 \sin 210°$$

$$- 0.4444 \sin 240°)$$

$$= -0.0576Fr$$

In a similar manner, we find $T_A = 0.997Fr$.   ////

The task of finding the deflection at any point on a ring with a loading like that of Fig. 5-3 is indeed difficult. The problem can be set up using Eq. (5-2), but the resulting integrals will be lengthy. The chances of making an error in signs or in terms during any of the simplification processes is very great. If a computer or even a programmable calculator is available, the integration can be performed using a numerical procedure such as Simpson's rule. Most of the user's manuals for programmable calculators contain such programs in the master library. When this approach is taken, the two terms behind each integral should not be multiplied out or simplified; reserve these tasks for the computer.

5-4-1   A Ring with Symmetrical Loads

A ring having three equally spaced loads, all equal in magnitude, with three equally spaced supports located midway betwen each pair of loads has reactions at each support of $R = F/2$, $M = 0.289Fr$, and $T = 0$ by Biezeno's theorem. To find the moment and torque at any location θ from a reaction, we construct the diagram shown in Fig. 5-5. Then the moment and torque at A are

$$M = M_1 \cos \theta - R_1 r \sin \theta$$
$$= Fr (0.289 \cos \theta - 0.5 \sin \theta) \tag{5-11}$$
$$T = M_1 \sin \theta - R_1 r (1 - \cos \theta)$$
$$= Fr (0.289 \sin \theta - 0.5 + 0.5 \cos \theta) \tag{5-12}$$

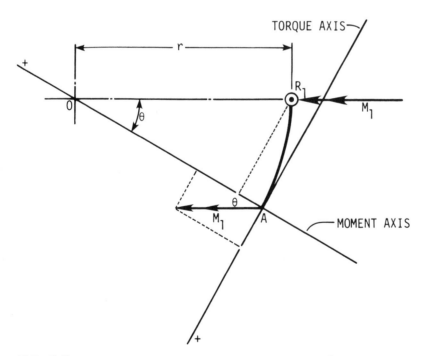

FIG. 5-5   The positive directions of the moment and torque axes are arbitrary. Note that $R_1 = F/2$ and $M_1 = 0.289Fr$.

Neglecting direct shear, the strain energy stored in the ring between any two supports is, from Table 5-2,

$$U = 2 \int_0^{\pi/3} \frac{M^2 r \, d\theta}{2EI} + 2 \int_0^{\pi/3} \frac{T^2 r \, d\theta}{2GK} \tag{5-13}$$

Castigliano's theorem states that the deflection at the load F is

$$y = \frac{\partial U}{\partial F} = \frac{2r}{EI} \int_0^{\pi/3} M \frac{\partial M}{\partial F} \, d\theta + \frac{2r}{GK} \int_0^{\pi/3} T \frac{\partial T}{\partial F} \, d\theta \tag{5-14}$$

From Eqs. (5-11) and (5-12) we find

$$\frac{\partial M}{\partial F} = r(0.289 \cos \theta - 0.5 \sin \theta)$$

$$\frac{\partial T}{\partial F} = r(0.289 \sin \theta - 0.5 + 0.5 \cos \theta)$$

When these are substituted into Eq. (5-14) we get

$$y = \frac{Fr^3}{2} \left(\frac{A}{EI} + \frac{B}{GK} \right) \tag{5-15}$$

which is the same as Eq. (5-5). The constants are

$$A = 4 \int_0^{\pi/3} (0.289 \cos \theta - 0.5 \sin \theta)^2 \, d\theta$$
$$B = 4 \int_0^{\pi/3} (0.289 \sin \theta - 0.5 + 0.5 \cos \theta)^2 \, d\theta \tag{5-16}$$

These equations can be integrated directly or by a computer using Simpson's rule. If your integration is rusty, use the computer. The results are $A = 0.1208$ and $B = 0.0134$.

5-4-2 Distributed Loading

The ring segment in Fig. 5-6 is subjected to a distributed load w per unit circumference and is supported by the vertical reactions R_1 and R_2 and the moment reactions M_1 and M_2. The zero-torque reactions mean that the ring is free to turn at A and B. The resultant of the distributed load is $W = wr\phi$; it acts at the centroid:

$$\bar{r} = \frac{2r \sin (\phi/2)}{\phi} \tag{5-17}$$

By symmetry, the force reactions are $R_1 = R_2 = W/2 = wr\phi/2$. Summing moments about an axis through BO gives

$$\Sigma M(BO) = -M_2 + W\bar{r} \sin \frac{\phi}{2} - M_1 \cos (\pi - \phi) - \frac{wr^2\phi}{2} \sin \phi = 0$$

Since M_1 and M_2 are equal, this equation can be solved to give

$$M_1 = wr^2 \left[\frac{1 - \cos \phi - (\phi/2) \sin \phi}{1 - \cos \phi} \right] \tag{5-18}$$

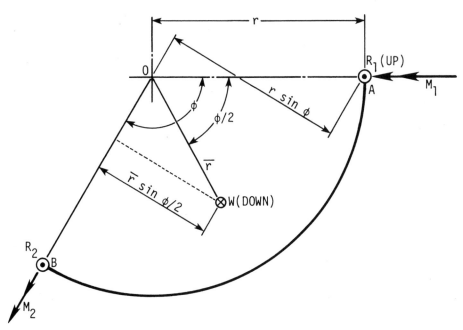

FIG. 5-6 Section of ring of span angle ϕ with distributed load.

EXAMPLE 2. A ring has a uniformly distributed load and is supported by three equally spaced reactions. Find the deflection midway between supports.

Solution. By placing a load Q midway between supports and computing the strain energy using half the span, Eq. (5-7) becomes

$$y = \frac{\partial U}{\partial Q} = \frac{2r}{EI} \int_0^{\phi/2} M \frac{\partial M}{\partial Q} \, d\theta + \frac{2r}{GK} \int_0^{\phi/2} T \frac{\partial T}{\partial Q} \, d\theta \qquad (5\text{-}19)$$

Using Eq. (5-18) with $\phi = 2\pi/3$ gives the moment at a support due only to w to be $M_1 = 0.395wr^2$. Then using a procedure quite similar to that used to write Eqs. (5-11) and (5-12) we find the moment and torque due only to the distributed load and at any section θ to be

$$M_w = wr^2 \left(1 - 0.605 \cos\theta - \frac{\pi}{3} \sin\theta \right)$$
$$T_w = wr^2 \left(\theta - 0.605 \sin\theta - \frac{\pi}{3} + \frac{\pi}{3} \cos\theta \right) \qquad (5\text{-}20)$$

In a similar manner, the force Q results in additional components of

$$M_Q = \frac{Qr}{2} (0.866 \cos\theta - \sin\theta)$$
$$T_Q = \frac{Qr}{2} (0.866 \sin\theta - 1 + \cos\theta) \qquad (5\text{-}21)$$

Then

$$\frac{\partial M_Q}{\partial Q} = \frac{r}{2} (0.866 \cos\theta - \sin\theta)$$

$$\frac{\partial T_Q}{\partial Q} = \frac{r}{2} (0.866 \sin\theta - 1 + \cos\theta)$$

Notes · Drawings · Ideas

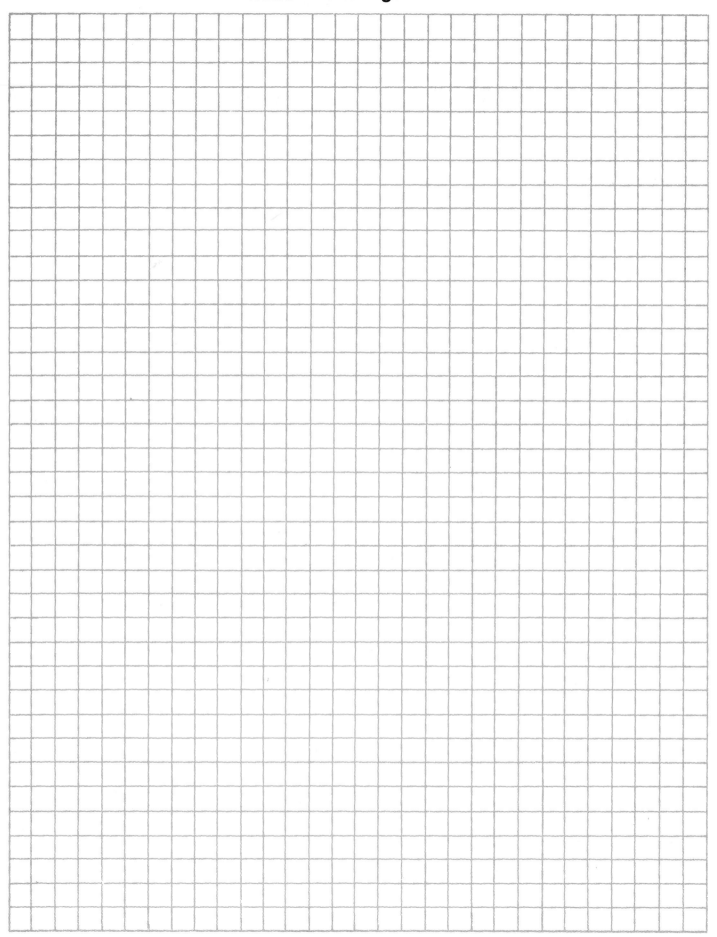

And so, placing the fictitious force Q equal to zero, Eq. (5-19) becomes

$$
y = \frac{wr^4}{EI} \int_0^{\pi/3} \left(1 - 0.605 \cos\theta - \frac{\pi}{3} \sin\theta \right) (0.866 \cos\theta - \sin\theta) \, d\theta
$$

$$
+ \frac{wr^4}{GK} \int_0^{\pi/3} \left(\theta - 0.605 \sin\theta - \frac{\pi}{3} + \frac{\pi}{3} \cos\theta \right) (0.866 \sin\theta - 1 + \cos\theta) \, d\theta \quad (5\text{-}22)
$$

When this expression is integrated, we find

$$
y = \frac{wr^4}{2} \left(\frac{0.141}{EI} + \frac{0.029}{GK} \right) \quad (5\text{-}23)
$$

$$////$$

5-5   RING SEGMENTS WITH FIXED ENDS

A ring segment with fixed ends has a moment reaction M_1, a torque reaction T_1, and a shear reaction R_1, as shown in Fig. 5-7a. The system is indeterminate, so all three relations of Eq. (5-2) must be used to determine them, using zero for each corresponding displacement.

5-5-1   Segment with Concentrated Load

The moment and torque at any position θ are found from Fig. 5-7b as

$$
M = T_1 \sin\theta + M_1 \cos\theta - R_1 r \sin\theta + Fr \sin(\theta - \gamma)
$$

$$
T = -T_1 \cos\theta + M_1 \sin\theta - R_1 r (1 - \cos\theta) + Fr[1 - \cos(\theta - \gamma)]
$$

These can be simplified; the result is

$$
M = T_1 \sin\theta + M_1 \cos\theta - R_1 r \sin\theta + Fr \cos\gamma \sin\theta - Fr \sin\gamma \cos\theta \quad (5\text{-}24)
$$

$$
T = -T_1 \cos\theta + M_1 \sin\theta - R_1 r (1 - \cos\theta)
$$

$$
- Fr \cos\gamma \cos\theta - Fr \sin\gamma \sin\theta + Fr \quad (5\text{-}25)
$$

Using Eq. (5-3) and the third relation of Eq. (5-2) gives

$$
\frac{\partial U}{\partial M_1} = \frac{r}{EI} \int_0^\phi M \frac{\partial M}{\partial M_1} \, d\theta + \frac{r}{GK} \int_0^\phi T \frac{\partial T}{\partial M_1} \, d\theta = 0 \quad (5\text{-}26)
$$

Note that

$$
\frac{\partial M}{\partial M_1} = \cos\theta
$$

$$
\frac{\partial T}{\partial M_1} = \sin\theta
$$

Now multiply Eq. (5-26) by EI and divide by r; then substitute. The result can be

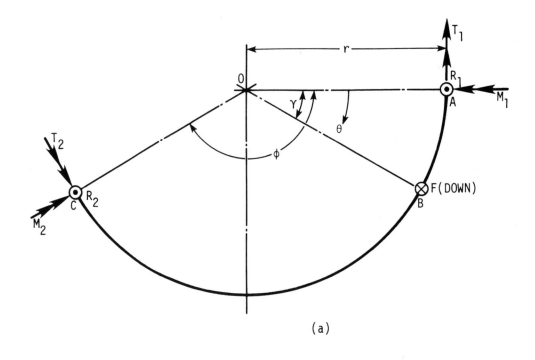

(a)

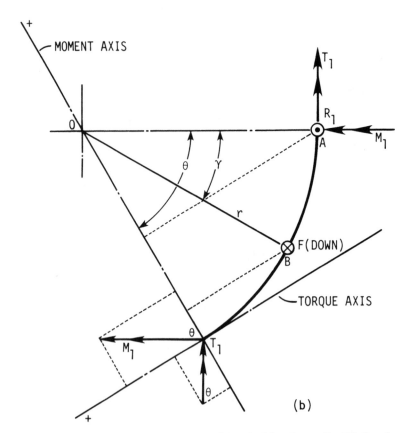

(b)

FIG. 5-7 (*a*) Ring segment of span angle ϕ loaded by force *F*. (*b*) Portion of ring used to compute moment and torque at position θ.

written in the form

$$\int_0^\phi (T_1 \sin \theta + M_1 \cos \theta - R_1 r \sin \theta) \cos \theta \, d\theta$$

$$+ Fr \int_\gamma^\phi (\cos \gamma \sin \theta - \sin \gamma \cos \theta) \cos \theta \, d\theta$$

$$+ \frac{EI}{GK} \left\{ \int_0^\phi [-T_1 \cos \theta + M_1 \sin \theta - R_1 r(1 - \cos \theta)] \sin \theta \, d\theta \right.$$

$$\left. - Fr \int_\gamma^\phi (\cos \gamma \cos \theta + \sin \gamma \sin \theta - 1) \sin \theta \, d\theta \right\} = 0 \quad (5\text{-}27)$$

Similar equations can be written using the other two relations in Eq. (5-2). When these three relations are integrated, the results can be expressed in the form

$$\begin{bmatrix} a_{11} & a_{12} & a_{13} \\ a_{21} & a_{22} & a_{23} \\ a_{31} & a_{32} & a_{33} \end{bmatrix} \begin{bmatrix} T_1/Fr \\ M_1/Fr \\ R_1/F \end{bmatrix} = \begin{bmatrix} b_1 \\ b_2 \\ b_3 \end{bmatrix} \quad (5\text{-}28)$$

where

$$a_{11} = \sin^2 \phi - \frac{EI}{GK} \sin^2 \phi \tag{5-29}$$

$$a_{21} = (\phi - \sin \phi \cos \phi) + \frac{EI}{GK} (\phi + \sin \phi \cos \phi) \tag{5-30}$$

$$a_{31} = (\phi - \sin \phi \cos \phi) + \frac{EI}{GK} (\phi + \sin \phi \cos \phi - 2 \sin \phi) \tag{5-31}$$

$$a_{12} = (\phi + \sin \phi \cos \phi) + \frac{EI}{GK} (\phi - \sin \phi \cos \phi) \tag{5-32}$$

$$a_{22} = a_{11} \tag{5-33}$$

$$a_{32} = \sin^2 \phi + \frac{EI}{GK} [2(1 - \cos \phi) - \sin^2 \phi] \tag{5-34}$$

$$a_{13} = -a_{32} \tag{5-35}$$

$$a_{23} = -a_{31} \tag{5-36}$$

$$a_{33} = -(\phi - \sin \phi \cos \phi) - \frac{EI}{GK} (3\phi - 4 \sin \phi + \sin \phi \cos \phi) \tag{5-37}$$

$$b_1 = \sin \gamma \sin \phi \cos \phi - \cos \gamma \sin^2 \phi + (\phi - \gamma) \sin \gamma + \frac{EI}{GK} (\cos \gamma \sin^2 \phi$$

$$- \sin \gamma \sin \phi \cos \phi + (\phi - \gamma) \sin \gamma + 2 \cos \phi - 2 \cos \gamma) \tag{5-38}$$

$$b_2 = (\gamma - \phi) \cos \gamma - \sin \gamma + \cos \gamma \sin \phi \cos \phi + \sin \gamma \sin^2 \phi$$

$$+ \frac{EI}{GK} [(\gamma - \phi) \cos \gamma - \sin \gamma + 2 \sin \phi - \cos \gamma \sin \phi \cos \phi - \sin \gamma \sin^2 \phi] \tag{5-39}$$

$$b_3 = (\gamma - \phi) \cos \gamma - \sin \gamma + \cos \gamma \sin \phi \cos \phi + \sin \gamma \sin^2 \phi)$$

$$+ \frac{EI}{GK} [(\gamma - \phi) \cos \gamma - \sin \gamma - \cos \gamma \sin \phi \cos \phi - \sin \gamma \sin^2 \phi$$

$$+ 2(\sin \phi - \phi + \gamma + \cos \gamma \sin \phi - \sin \gamma \cos \phi)] \tag{5-40}$$

TABLE 5-4 Coefficients a_{ij} for Various Span Angles

Coefficients		Span angle ϕ						
		$3\pi/2$	π	$3\pi/4$	$2\pi/3$	$\pi/2$	$\pi/3$	$\pi/4$
a_{11}	X_{11}	1	0	0.5	0.75	1	0.75	0.5
	Y_{11}	-1	0	-0.5	-0.75	-1	-0.75	-0.5
a_{21}	X_{21}	4.7124	π	2.8562	2.5274	1.5708	0.6142	0.2854
	Y_{21}	4.7124	π	1.8562	1.6614	1.5708	1.4802	1.2854
a_{31}	X_{31}	4.7124	π	2.8562	2.5274	1.5708	0.6142	0.2854
	Y_{31}	6.7124	π	0.4420	-0.0707	-0.4292	-0.2518	-0.1288
a_{12}	X_{12}	4.7124	π	1.8562	1.6614	1.5708	1.4802	1.2854
	Y_{12}	4.7124	π	2.8562	2.5274	1.5708	0.6142	0.2854
a_{22}	X_{22}	1	0	0.5	0.75	1	0.75	0.5
	Y_{22}	-1	0	-0.5	-0.75	-1	-0.75	-0.5
a_{32}	X_{32}	1	0	0.5	0.75	1	0.75	0.5
	Y_{32}	1	4	2.9142	2.25	1	0.25	0.0858
a_{13}	X_{13}	-1	0	-0.5	-0.75	-1	-0.75	-0.5
	Y_{13}	-1	-4	-2.9142	-2.25	-1	-0.25	-0.0858
a_{23}	X_{23}	-4.7124	$-\pi$	-2.8562	-2.5274	-1.5708	-0.6142	-0.2854
	Y_{23}	-6.7124	$-\pi$	-0.4420	0.0707	0.4292	0.2518	0.1288
a_{33}	X_{33}	-4.7124	$-\pi$	-2.8562	-2.5274	-1.5708	-0.6142	-0.2854
	Y_{33}	-18.1372	-3π	-3.7402	-2.3861	-0.7124	-0.1105	-0.0277

For tabulation purposes we indicate these relations in the form

$$a_{ij} = X_{ij} + \frac{EI}{GK} Y_{ij} \qquad b_k = X_k + \frac{EI}{GK} Y_k \tag{5-41}$$

Programs for solving equations such as Eq. (5-28) are widely available and easy to use. Tables 5-4 and 5-5 list the values of the coefficients for a variety of span and load angles.

5-5-2 Deflection Due to Concentrated Load

The deflection of a ring segment at a concentrated load can be obtained using the first relation of Eq. (5-2). The complete analytical solution is quite lengthy, so a result is shown here that can be solved using computer solutions of Simpson's approximation. First, define the three solutions to Eq. (5-28) as

$$T_1 = C_1 Fr \qquad M_1 = C_2 Fr \qquad R_1 = C_3 F \tag{5-42}$$

Then Eq. (5-2) will have four integrals, which are

$$A_F = \int_0^\phi [(C_1 - C_3) \sin \theta + C_2 \cos \theta]^2 \, d\theta \tag{5-43}$$

$$B_F = \int_0^\phi (\cos \gamma \sin \theta - \sin \gamma \cos \theta)^2 \, d\theta \tag{5-44}$$

TABLE 5-5 Coefficients b_k for Various Span Angles ϕ and Load Angles γ in Terms of ϕ

Coefficients, load angles γ			Span angle ϕ						
			$3\pi/2$	π	$3\pi/4$	$2\pi/3$	$\pi/2$	$\pi/3$	$\pi/4$
$\dfrac{\phi}{4}$	b_1	X_1	2.8826	1.6661	0.2883	-0.0806	-0.4730	-0.4091	-0.2780
		Y_1	2.8826	-1.7481	-1.4019	-1.0806	-0.4730	-0.1162	-0.0396
	b_2	X_2	-1.3525	-2.3732	-2.1628	-1.8603	-1.0884	-0.4051	-0.1849
		Y_2	-5.2003	-2.3732	-0.4727	-0.1283	0.1462	0.1022	0.0535
	b_3	X_3	-1.3525	-2.3732	-2.1628	-1.8603	-1.0884	-0.4051	-0.1849
		Y_3	-13.0342	-5.6714	-2.0455	-1.2699	-0.3622	-0.0544	-0.0135
$\dfrac{\phi}{3}$	b_1	X_1	3.1416	1.8138	0.4036	0.0446	-0.3424	-0.3179	-0.2180
		Y_1	3.1416	-1.1862	-1.0106	-0.7817	-0.3424	-0.0839	-0.0286
	b_2	X_2	0	-1.9132	-1.8178	-1.5620	-0.9069	-0.3346	-0.1522
		Y_2	-4	-1.9132	-0.4036	-0.1307	0.0931	0.0706	0.0373
	b_3	X_3	0	-1.9132	-1.8178	-1.5620	-0.9069	-0.3346	-0.1522
		Y_3	-10.2832	-4.3700	-1.5452	-0.9536	-0.2692	-0.0401	-0.0099
$\dfrac{\phi}{2}$	b_1	X_1	2.3732	1.5708	0.4351	0.1569	-0.1517	-0.1712	-0.1203
		Y_1	2.3732	-0.4292	-0.4379	-0.3431	-0.1517	-0.0372	-0.0127
	b_2	X_2	1.6661	-1	-1.1041	-0.9566	-0.5554	-0.2034	-0.0922
		Y_2	-1.7481	-1	-0.2311	-0.0906	0.0304	0.0286	0.0154
	b_3	X_3	1.6661	-1	-1.1041	-0.9566	-0.5554	-0.2034	-0.0922
		Y_3	-5.0463	-2.1416	-0.7395	-0.4529	-0.1262	-0.0186	-0.0046

$$C_F = \int_0^\phi [(C_3 - C_1) \cos \theta + C_2 \sin \theta - C_3]^2 \, d\theta \qquad (5\text{-}45)$$

$$D_F = \int_0^\phi [1 - (\cos \gamma \cos \theta + \sin \gamma \sin \theta)]^2 \, d\theta \qquad (5\text{-}46)$$

The results of these four integrations should be substituted into

$$y = \frac{Fr^3}{EI} \left[A_F + B_F + \frac{EI}{GK}(C_F + D_F) \right] \qquad (5\text{-}47)$$

to obtain the deflection due to F and at the location of the force F.

It is worth noting that the point of maximum deflection will never be far from the middle of the ring, even though the force F may be exerted near one end. This means that Eq. (5-47) will not give the maximum deflection unless $\gamma = \phi/2$.

5-5-3 Segment with Distributed Load

The resultant load acting at the centroid B' in Fig. 5-8 is $W = wr\phi$, and the radius $\bar{r}$ is given by Eq. (5-7), with ϕ substituted for θ. Thus the shear reaction at the fixed end A is $R_1 = wr\phi/2$. M_1 and T_1, at the fixed ends, can be determined using Castigliano's method.

We use Fig. 5-9 to write equations for moment and torque for any section, such as the one at D. When Eq. (5-6) for $\bar{r}$ is used, the results are found to be

$$M = T_1 \sin \theta + M_1 \cos \theta - \frac{wr^2\phi}{2} \sin \theta + wr^2(1 - \cos \theta) \qquad (5\text{-}48)$$

$$T = -T_1 \cos \theta + M_1 \sin \theta - \frac{wr^2\phi}{2}(1 - \cos \theta) + wr^2(\theta - \sin \theta) \qquad (5\text{-}49)$$

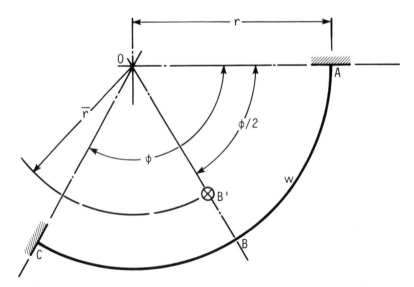

FIG. 5-8 Ring segment of span angle ϕ subjected to a uniformly distributed load w per unit circumference acting downward. Point B' is the centroid of the load. The ends are fixed to resist bending moment and torsional moment.

These equations are now employed in the same manner as in Sec. 5-5-1 to obtain

$$\begin{bmatrix} a_{11} & a_{12} \\ a_{21} & a_{22} \end{bmatrix} \begin{bmatrix} T_1/wr^2 \\ M_1/wr^2 \end{bmatrix} = \begin{bmatrix} b_1 \\ b_2 \end{bmatrix} \tag{5-50}$$

It turns out that the array containing the a_{ij} terms is identical with the same coefficients in Eq. (5-28); they are given by Eqs. (5-29), (5-30), (5-32), and (5-33), respectively. The coefficients b_k are

$$b_k = X_k + \frac{EI}{GK} Y_k \tag{5-51}$$

where

$$X_1 = \frac{\phi}{2} \sin^2 \phi + \sin \phi \cos \phi + \phi - 2 \sin \phi \tag{5-52}$$

$$Y_1 = \phi - 2 \sin \phi - \frac{\phi}{2} \sin^2 \phi - \sin \phi \cos \phi + \phi(1 + \cos \phi) \tag{5-53}$$

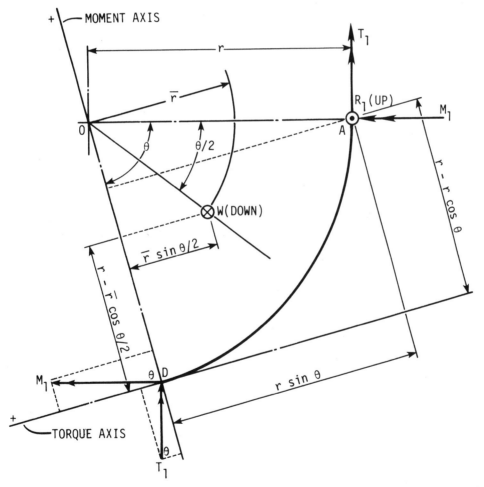

FIG. 5-9 A portion of the ring has been isolated here to determine the moment and torque at any section D at angle θ from the fixed end at A.

Notes ▪ Drawings ▪ Ideas

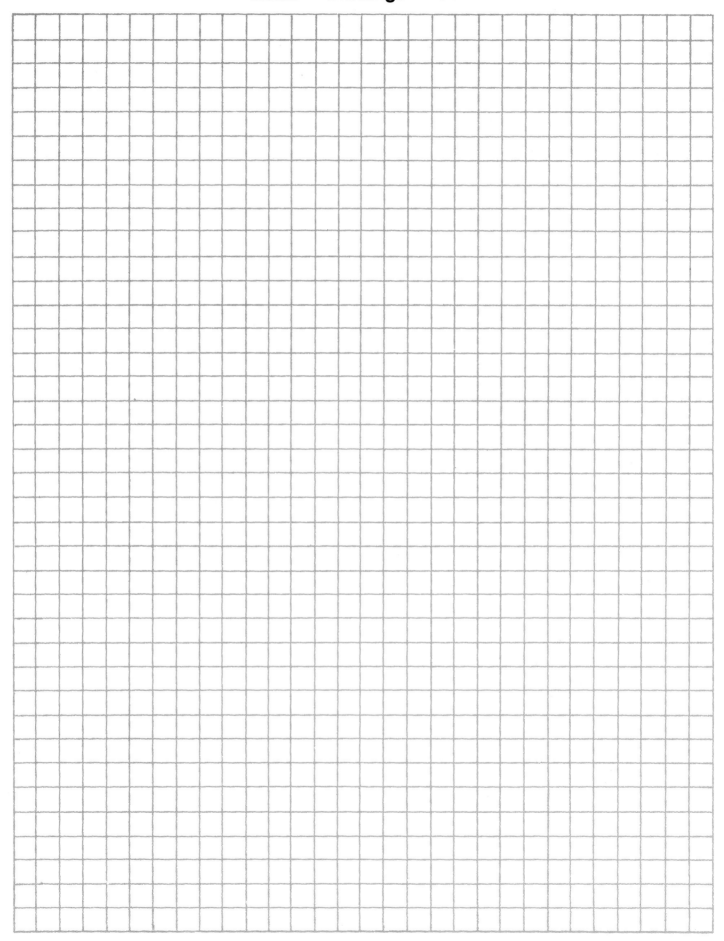

TABLE 5-6 Coefficients b_k for Various Span Angles and Uniform Loading

Coefficients		Span angle ϕ						
		$3\pi/2$	π	$3\pi/4$	$2\pi/3$	$\pi/2$	$\pi/3$	$\pi/4$
b_1	X_1	9.0686	3.1416	1.0310	0.7147	0.3562	0.1409	0.0675
	Y_1	9.0686	3.1416	1.5430	1.0572	0.3562	0.0602	0.0156
b_2	X_2	10.1033	0.9348	0.4507	0.3967	0.2337	0.0716	0.0263
	Y_2	8.1033	0.9348	−1.7274	−2.0102	−1.7663	−0.9750	−0.5810

$$X_2 = \frac{\phi^2}{2} - 2(1 - \cos\phi) - \frac{\phi}{2}\sin\phi\cos\phi + \sin^2\phi \tag{5-54}$$

$$Y_2 = \frac{\phi^2}{2} - 2(1 - \cos\phi) + \frac{\phi}{2}\sin\phi\cos\phi - \sin^2\phi + \phi\sin\phi \tag{5-55}$$

Solutions to these equations for a variety of span angles are given in Table 5-6.

A solution for the deflection at any point can be obtained using a fictitious load Q at any point and proceeding in a manner similar to other developments in this chapter. It is, however, a very lengthy analysis.

REFERENCES

5-1 Warren C. Young, *Roark's Formulas for Stress and Strain*, 6th ed., McGraw-Hill, 1989.

5-2 Joseph E. Shigley and Larry D. Mitchell, *Mechanical Engineering Design*, 4th ed., McGraw-Hill, 1983.

5-3 J. P. Den Hartog, *Advanced Strength of Materials*, McGraw-Hill, 1952.

Notes • Drawings • Ideas

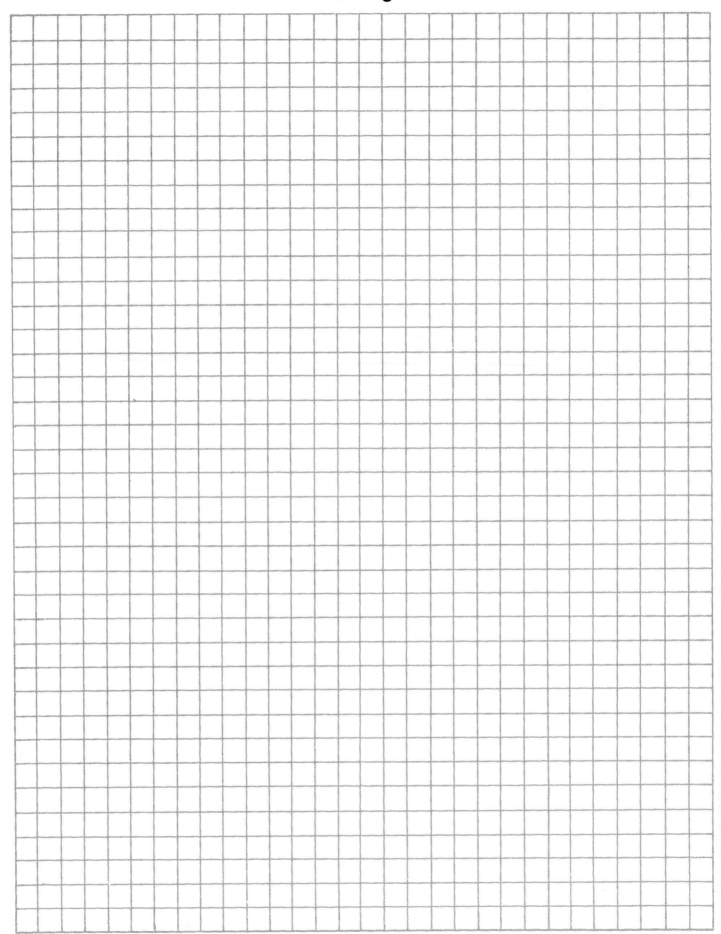

chapter *6*
PRESSURE CYLINDERS

SACHINDRANARAYAN BHADURI, Ph.D.
Associate Professor
Mechanical and Industrial Engineering Department
The University of Texas at El Paso
El Paso, Texas

6-1   INTRODUCTION

The pressure vessels commonly used in industrial applications consist basically of a few closed shells of simple shape: spherical or cylindrical with hemispherical, conical, ellipsoidal, or flat ends. The shell components are joined together mostly by welding and riveting; sometimes they are bolted together using flanges.

Generally, the shell elements are axisymmetrical surfaces of revolution formed by rotation of a straight line or a plane curve known as a *meridian* or a *generator* about an axis of rotation. The plane containing the axis of rotation is called the *meridional plane*. The geometry of such simple shells is specified by the form of the midwall surface, usually two radii of curvature and the wall thickness at every point. The majority of pressure vessels are cylindrical.

In practice, the shell is considered thin if the wall thickness t is small in comparison with the circumferential radius of curvature R_θ and the longitudinal radius of curvature R_ℓ. If the ratio $R_\theta/t > 10$, the shell is considered to be thin shell. This implies that the stresses developed in the shell wall by external loads can be considered to be uniformly distributed over the wall thickness. Many shells used in pressure-vessel construction are relatively thin ($10 < R_\theta/t < 500$), with the associated uniform distribution of stresses throughout the cylinder wall. Bending stresses in the walls of such membrane shells due to concentrated external loads are of higher intensity near the area of application of the load. The attenuation distance from the load where the stresses die out is short. The radial deformation of a shell subjected to internal pressure is assumed smaller than one-half the shell thickness. The shell thickness is designed to keep the maximum stresses below the yield strength of the material.

6-2   DESIGN PRINCIPLES OF PRESSURE CYLINDERS

In the design of a pressure vessel as a unit, a number of criteria should be considered. These are (1) selection of the material for construction of the vessel based on a working knowledge of the properties of the material, (2) determination of the magnitude of the induced stress in conformity with the requirements of standard codes, and (3) determination of the elastic stability. To simplify the design and keep the cost of fabrication low, the components of a vessel should be made in the form of simple geometric shapes, such as spherical, cylindrical, or conical. A spherical geometry provides minimum surface area per unit volume and requires minimum wall thickness for a given pressure. From the point of view of material savings and uniform distribution of induced stresses in the shell wall, a spherical shape is favorable. However, the fabrication of spherical vessels is more complicated and expensive than that of cylindrical ones. Spherical vessels are used commonly for storage of gas and liquids. For large-volume, low-pressure storage, spherical vessels are economical. But for higher-pressure storage, cylindrical vessels are more economical.

The most common types of vessels can be classified according to their

1. Functions: Storage vessels, reactors, boilers, mixers, and heat exchangers
2. Structural materials: Steel, cast iron, copper, plastics, etc.
3. Method of fabrication: Welded, cast, brazed, flanged, etc.
4. Geometry: Cylindrical, spherical, conical, combined
5. Scheme of loading: Working under internal or external pressure
6. Wall temperatures: Heated, unheated
7. Corrosion action: Moderate or high corrosion effects
8. Orientation in space: Vertical, horizontal, sloped
9. Method of assembly: Detachable, nondetachable
10. Wall thickness: Thin-walled ($d_o/d_i < 1.5$); thick-walled ($d_o/d_i \geq 1.5$)

Most vessels are designed as cylindrical shells fabricated of rolled sheet metal or as cylindrical shells that are cast. From the point of view of simplified structural design, the stressed state of the material of a thin-wall shell is considered biaxial. This is permissible because the magnitude of the radial stress in such a vessel wall is very small. The stressed state of the shell wall is generally the sum of the two basic components: (1) stressed state due to uniformly distributed forces on the surface as a result of fluid pressure, and (2) stressed state due to the action of the forces and moments distributed around the contour.

The stressed condition due to uniform pressure of fluid on the surface of the shell can be determined either by the membrane or moment theory. The *membrane theory* yields accurate enough results for most engineering applications and is widely used for structural designs. The *moment theory* is not usually applied for the determination of stresses due to uniformly distributed fluid forces on a surface. The equations resulting from the application of the moment theory are complex and the design process is quite involved.

End forces and moments are calculated in sections where a sudden change in load, wall thickness, or the properties of the shell material occurs. The stressed state induced by the applied end forces and moments are determined by application of the moment theory [6-1]. The induced stress and deformations due to end effects influence mostly the zones where the end forces and moments are applied. In general, end stresses should be carefully evaluated, and design measures must be taken to keep them within the safe limit. High values of end stresses are to be avoided, especially for brittle materials and vessels operating under high alternating loads.

The structural reliability of the equipment parts is determined by two different approaches. The *theory of elasticity* requires that the strength be determined by the ultimate stress which the part can withstand without rupture, whereas the *theory of plasticity* suggests that the strength be determined by the ultimate load the part can withstand without residual deformation. The elastic theory is based on the assumption that the material of the component parts of the vessel is in an elastic state everywhere and nowhere the state of stress exceeds the yield point.

The parts of a pressure vessel are not generally bonded uniformly. When the

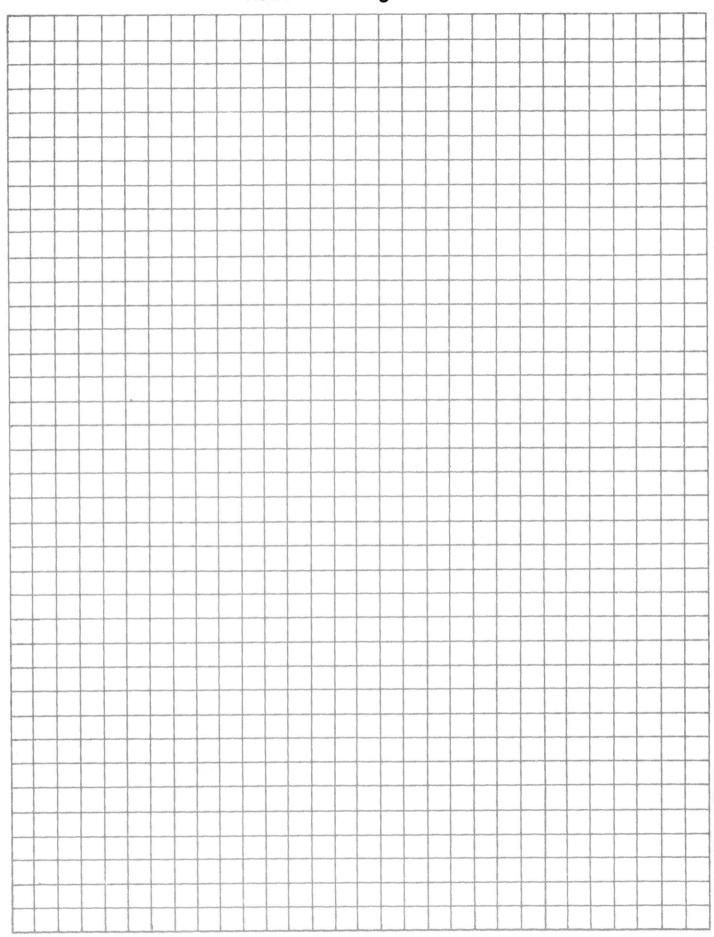

structural design is based on the ultimate stress occurring at the most loaded region of the construction, the required material consumption could be excessive. In design works using the membrane theory, only average stresses are assumed, and no efforts are made to include the local stresses of significant magnitude. However, the designer should consider the probable adverse effects of very high local stress intensities and modify the design accordingly.

In heavily loaded parts of pressure vessels made of plastic materials, partial transition to the elastic-plastic state occurs. The plastic design method permits a realistic evaluation of the maximum load this vessel can withstand without failure.

6-3 DESIGN LOADS

The principal loads (i.e., forces) applied in actual operations to a vessel or its structural attachments to be considered in the design of such a vessel are

1. Design pressure (internal or external or both)
2. Dead loads
3. Wind loads
4. Thermal loads
5. Piping load
6. Impact load
7. Cyclic loads

Several loading combinations are often possible. The designer should consider them carefully for a particular situation and select the most probable combination of simultaneous loads for an economical and reliable design. Failure of a pressure vessel may be due to improper selection of materials, defects in materials due to inadequate quality control, incorrect design conditions, computational errors in design, improper or faulty fabrication procedures, or inadequate inspection and shop testing.

Design pressure is the pressure used to determine the minimum thickness of each of the vessel components. It is the difference between the internal and external pressures. A suitable margin above the operating pressure (usually 10 percent of operating pressure and a minimum of 10 psi) plus a static head of operating liquid must be included. The minimum design pressure for a "code" nonvacuum vessel is 15 psi. Vessels with negative gauge operating pressures are designed for full vacuum. In determining the design load, the designer must consult the *ASME Boiler and Pressure Vessel Code,* Sec. VIII, Pressure Vessels, Division I. The maximum operating pressure is, according to code definition, the maximum gauge pressure permissible at the top of the completed vessel in its operating condition at the designated temperature. It is based on the nominal thickness, exclusive of the corrosion allowance, and the thickness required for loads other than fluid pressure.

The *required thickness* is the minimum thickness of the vessel wall, computed by code formula. The *design thickness* is the minimum required thickness plus an allowance for corrosion. The *nominal thickness* is the design thickness of the commercially available material actually used in making the vessel.

The vessel shell must be designed to withstand the combined effect of the pressure and temperature under the anticipated operating conditions. One has to use the pressure-vessel code and standard engineering analysis whenever applicable to determine

Notes ▪ Drawings ▪ Ideas

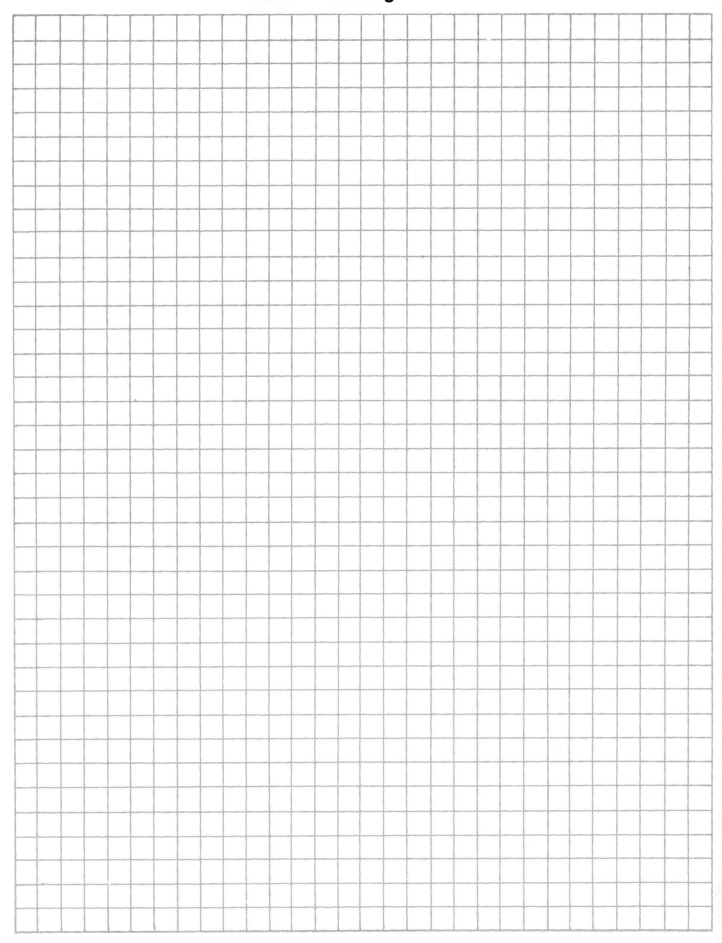

the nominal stress at any part of the vessel. The stresses cannot in general exceed the allowable stress obtained by the code guidelines.

Design temperature is really more a design environmental condition than a load. *Thermal loads* originate from temperature changes combined with body restraints or existing temperature gradients. Reduction in structural strength due to rising temperature, increase in brittleness with rapidly falling temperature, and the associated changes in the vessel dimensions should be taken into careful consideration for the design. The code design temperature must not be less than the mean vessel-wall temperature under operating conditions. It can be determined by the standard heat-transfer methods. Generally, the standard vessel design temperature is equal to the maximum operating temperature of the fluid in the vessel plus 50°F for safety considerations. For low-temperature operation, the minimum fluid temperature is used for the design temperature. The designer should consult the *ASME Boiler and Pressure Vessel Codes* and the excellent books by Bednar [6-2] and Chuse [6-3].

Dead loads are the forces due to the weight of the vessel and the parts permanently connected with the vessel. The *piping loads* acting on the vessel must be evaluated carefully. They consist of the weight of the pipe sections supported by the nozzles into the vessel shell, as well as operating forces due to thermal expansion of the pipes.

If the pressure vessel is exposed to the environment, the dynamic pressure forces due to the turbulent flow of air and associated gustiness must be evaluated for structural stability of the vessel support system. Determination of *wind load* is outlined in the ANSI A58.1-1972 code, and the designer must follow the guidelines to determine the effective dynamic load due to wind speed past the vessel and support-structure system. The same code provides guidelines for evaluation of seismic loads on flexible tall vessels.

In actual design situations, many combinations of loads are possible. Consequently, the designer should use discretion to determine the important ones and select only certain sets of design loads which can most probably occur simultaneously.

6-4 CYLINDRICAL SHELLS—STRESS ANALYSIS

Cylindrical shells are widely used in the manufacture of pressure vessels. They can be easily fabricated and have great structural strength. Depending on their function, they may be vertical or horizontal. Vertical pressure vessels are often preferred, especially for a thin-walled vessel operating under low internal pressure. The design of a vertical cylindrical vessel becomes simple because the additional bending stresses due to weight of the vessel itself and of the fluid can be eliminated.

Cylindrical shells of malleable materials, such as steels, nonferrous metals, and their alloys, operating under internal pressures up to 10 MPa are fabricated mostly of rolled and welded sheets of corrosion-resistant material (see Ref. [6-1]). The minimum thickness of a shell rolled from low-carbon sheet metal and welded is 4.0 mm. The minimum thickness for an austenitic steel shell is 3.4 mm; for copper shells, 2.5 to 30 mm; and for cast shells, 20 to 25 mm.

6-4-1 Cylindrical Storage Vessel

Before the actual design calculation of a vertical cylindrical vessel is done, the relationship between the optimum height and the diameter should be determined. The volume of the sheet metal needed to make a vertical cylindrical storage tank (Fig. 6-1) is determined by the following formula:

$$V_s = \pi d_o H t_1 + \frac{\pi d_o^2}{4}(t_k + t_a)$$

$$= \pi d_o H t_1 + \frac{\pi d_o^2}{4} t_2 \tag{6-1}$$

Vessel capacity is

$$V = \frac{\pi d_i^2}{4} H \tag{6-2}$$

The inside diameter of the vessel is

$$d_i = \sqrt{\frac{4V}{\pi H}} \tag{6-3}$$

Since thickness t_1 is very small in comparison with d_o and d_i, we may consider $d_i \approx d_o$. Substitution of Eq. (6-3) into Eq. (6-1), gives

$$V_s = 2t_1 \sqrt{\pi V H} + \frac{V t_2}{H} \tag{6-4}$$

The miminum volume of sheet metal is obtained by differentiating Eq. (6-4) with respect to H and equating the derivative to zero. Thus

$$\frac{dV_s}{dH} = t_1 \sqrt{\pi V/H} - \frac{V t_2}{H^2} = 0 \tag{6-5}$$

The optimum height and the optimum diameter of the tank are given by

$$H_{opt} = \left[\frac{V}{\pi}\left(\frac{t_2}{t_1}\right)^2 \right]^{1/3} \tag{6-6}$$

and

$$d_{opt} = 2 \left[\frac{V t_1}{\pi t_2} \right]^{1/3} \tag{6-7}$$

The shell thickness is given by

$$t_1 = \frac{H_{opt} \gamma d_{opt}}{2\sigma_a} \tag{6-8}$$

where γ = specific weight of the fluid in the tank, and σ_a = allowable stress. Substituting the optimum values of height and outside diameter in Eq. (6-8), the tank capacity V is recovered as

$$V = \pi t_1^2 \sqrt{\frac{\sigma_a^3}{\gamma^3 t_2}} \tag{6-9}$$

Considering a minimum shell thickness of $t_1 = 4$ mm and top and bottom thicknesses of 4 mm, we find $t_2 = 8$ mm. The specific weight of the liquid can be obtained from appropriate property tables. The allowable stress for low-carbon steel is 137

MPa. Using this information in Eq. (6-9), one can easily find the corresponding tank volume.

Pressure vessels subjected to internal and/or external pressures require application of the theoretical principles involved in shell analysis. The structural configurations of relatively simple geometric shapes such as cylinders and spheres have been studied extensively in the theories of plates and shells (see Refs. [6-2], [6-4], and [6-5]. In fact, no single chapter or even an entire textbook can cover all the advancements in the field, particularly where internal pressure, external pressure, and other modes of loading are present. Therefore, attempts will be made here only to cover the materials which are pertinent for design of simple pressure vessels and piping. The majority of piping and vessel components are designed for internal pressure and have been analyzed to a great degree of sophistication. Numerous cases involve application of external pressure as well, where stresses, elastic stability, and possible structural failure must be analyzed and evaluated.

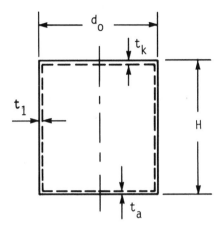

FIG. 6-1 Cylindrical storage vessel.

6-4-2 Membrane Theory

The common geometric shapes of pressure vessels used in industrial processes are spheres, cylinders, and ellipsoids. Conical and toroidal configurations are also used. Membrane theory can be applied to determine the stresses and deformations of such vessels when they have small thicknesses compared with other dimensions and have limited and small bending resistance normal to their surface. The stresses are considered to be average tension or compression over the thickness of the pipe or vessel wall acting tangential to the surface subjected to normal pressure. The imaginary surface passing through the middle of the wall thickness, however, extends in two directions and calls for rather complicated mathematical analysis, particularly when more than one expression for the curvature is necessary to describe the displacement of a point in the shell wall. In fact, in the more general sense it is necessary to define a normal force, two transverse shear forces, two bending moments, and a torque in order to evaluate the state of stress at a point. Membrane theory simplifies this analysis to a great extent and permits one to ignore bending and twisting moments when shell thickness is small. In many practical cases, consideration of equilibrium of the forces allows us to develop necessary relations for stresses and displacements in terms of the shell parameters for adequate design.

6-4-3 Thin Cylindrical Shells under Internal Pressure

When a thin cylinder is subjected to an internal pressure, three mutually perpendicular principal stresses—hoop stress, longitudinal stress, and radial stress—are developed in the cylinder material. If the ratio of thickness t and the inside diameter of the cylinder d_i is less than 1:20, membrane theory may be applied and we may assume that the hoop and longitudinal stresses are approximately constant across the wall thickness. The magnitude of radial stress is negligibly small and can be ignored. It is to be understood that this simplified approximation is used extensively for the design of thin cylindrical pressure vessels. However, in reality, radial stress varies from zero at the outside surface to a value equal to the internal pressure at the inside surface. The ends of the cylinder are assumed closed. Hoop stress is set up in resisting the bursting effect of the applied pressure and is treated by taking the equilibrium of half of the cylindrical vessel, as shown in Fig. 6-2. Total force acting on the half cylinder is

$$F_h = p_i d_i L \tag{6-10}$$

where d_i = inside diameter of cylinder, and L = length of cylinder. The resisting force due to hoop stress σ_h acting on the cylinder wall, for equilibrium, must equal the force F_h. Thus

$$F_h = 2\sigma_h t T \tag{6-11}$$

Substituting for F_h from Eq. (6-10) into Eq. (6-11), one obtains the following relation:

$$\sigma_h = \frac{p_i d_i}{2t} \quad \text{or} \quad \sigma_h = \frac{p_i r_i}{t} \tag{6-12}$$

Despite its simplicity, Eq. (6-12) has wide practical applications involving boiler drums, accumulators, piping, casing chemical processing vessels, and nuclear pres-

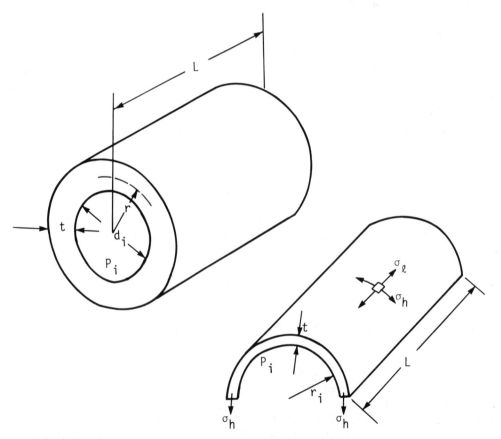

FIG. 6-2 Thin cylindrical shell under internal pressure.

sure vessels. Equation (6-12) gives the maximum tangential stress in the vessel wall on the assumption that the end closures do not provide any support, as is the case with long cylinders and pipes. Hoop stress can also be expressed in terms of the radius of the circle passing through the midpoint of the thickness. Then we can write

$$\sigma_h = \frac{p_i(r_i + 0.5t)}{t} \tag{6-13}$$

The shell thickness is then expressed as

$$t = \frac{p_i r_i}{\sigma_h - 0.5p_i} \tag{6-14}$$

The code stress and shell thickness formulas based on inside radius approximate the more accurate thick-wall formula of Lamé, which is

$$t = \frac{p_i r_i}{Se - 0.6p_i} \tag{6-15}$$

where e = code weld-joint efficiency, and S = allowable code stress.

Consideration of the equilibrium forces in the axial direction gives the longitudinal stress as

$$\sigma_\ell = \frac{p_i d_i}{4t}$$

or

$$\sigma_\ell = \frac{p_i r_i}{2t} \tag{6-16}$$

Equations (6-12) and Eqs. (6-16) reveal that the efficiency of the circumferential joint needs only be one-half that of the longitudinal joint. The preceding relations are good for elastic deformation only. The consequent changes in length, diameter, and intervolume of the cylindrical vessel subjected to inside fluid pressure can be determined easily.

The change in length is determined from the longitudinal strain ε_ℓ given by

$$\varepsilon_\ell = \frac{1}{E}(\sigma_\ell - \nu\sigma_h) \tag{6-17}$$

where E = modulus of elasticity, and ν = Poisson's ratio. The change in length ΔL is then given by

$$\Delta L = \frac{L}{E}(\sigma_\ell - \nu\sigma_h)$$

or

$$\text{or } \Delta L = \frac{p_i d_i}{4Et}(1 - 2\nu)L \tag{6-18}$$

Radial growth or dilatation under internal pressure is an important criterion in pipe and vessel analysis. For a long cylindrical vessel, the change in diameter is determined from consideration of the change in the circumference due to hoop stress. The change in circumference is obtained by multiplying hoop strain ε_h by the original circumference. The changed, or new, circumference is found to be equal $(\pi d_i + \pi d_i \varepsilon_h)$. This is the circumference of the circle of diameter $d_i(1 + \varepsilon_h)$. It can be shown

Notes ▪ Drawings ▪ Ideas

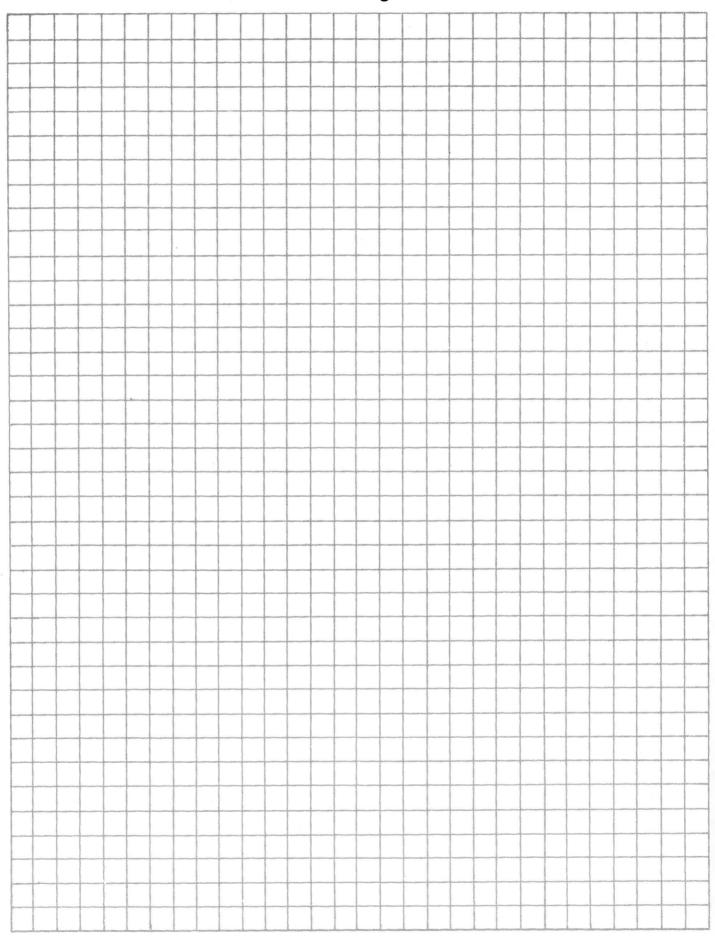

easily that the diametral strain equals the hoop or circumferential strain. The change in diameter Δd_i is given by

$$\Delta d_i = \frac{d_i}{E}(\sigma_h - \nu\sigma_\ell)$$

(6-19)

or

$$\Delta d_i = \frac{p_i d_i^2}{4Et}(2 - \nu)$$

The change in the internal volume ΔV of the cylindrical vessel is obtained by multiplying the volume strain by the original volume of the vessel and is given by

$$\Delta V = \frac{p_i d_i}{4Et}(5 - 4\nu)V_o$$

(6-20)

where ΔV = change in volume, and V_o = original volume.

6-4-4 Thin Spherical Shell under Internal Pressure

A sphere is a symmetrical body. The internal pressure in a thin spherical shell will set up two mutually perpendicular hoop stresses of equal magnitude and a radial stress. When the thickness-to-diameter ratio is less than 1:20, membrane theory permits us to ignore the radial stress component. The stress system then reduces to one of equal biaxial hoop or circumferential stresses.

Considering the equilibrium of the half sphere, it can be seen that the force on the half sphere (Fig. 6-3) due to internal pressure p_i is

$$F = \frac{\pi}{4}d_i^2 p_i$$

(6-21)

The resisting force due to hoop stress is given by

$$F_h = \pi d_i t \sigma_h$$

(6-22)

For equilibrium, the resistive force must be equal to the force due to pressure. Therefore,

$$\frac{\pi}{4}d_i^2 p_i = \pi d_i t \sigma_h$$

or

$$\sigma_h = \frac{p_i d_i}{4t}$$

(6-23)

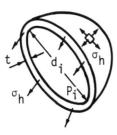

FIG. 6-3 Spherical shell under internal pressure.

Equation (6-23) gives the relevant maximum stress in a spherical shell. The expression for hoop stress for a thin cylindrical shell and that for a thin spherical shell are similar. This simple deduction is of great importance in the design of pressure vessels because the thickness requirement for a spherical vessel of the same material strength and thickness-to-diameter ratio is only one-half that required for a cylindrical shell.

The change in internal volume ΔV of a spherical shell can be evaluated easily from consideration of volumetric strain and the original volume. Volumetric strain is equal to the sum of three equal and mutually perpendicular strains. The change in internal volume due to internal pressure is given by

$$\Delta V = \frac{3p_i d_i}{4Et}(1 - \nu)V_o \qquad (6\text{-}24)$$

6-4-5 Vessels Subjected to Fluid Pressure

During the process of pressurization of a vessel, the fluid used as the medium changes in volume as the pressure is increased, and this must be taken into account when determining the amount of fluid which must be pumped into a cylinder in order to increase the pressure level in the vessel by a specified amount. The cylinder is considered initially full of fluid at atmospheric pressure. The necessary change in volume of the pressurizing fluid is given by

$$\Delta V_f = \frac{pV_o}{K} \qquad (6\text{-}25)$$

where K = bulk modulus of fluid
 p = pressure
 V_o = original volume

The additional amount of fluid necessary to raise the pressure must take up the change in volume given by Eq. (6-25) together with the increase in internal volume of the cylinder. The amount of additional fluid V_a required to raise the cylinder pressure by p is given by

$$V_a = \frac{pd_i}{4Et}[5 - 4\nu]V + \frac{pV}{K} \qquad (6\text{-}26)$$

It can be shown by similar analysis that the additional volume of fluid V_a required to pressurize a spherical vessel is given by

$$V_a = \frac{3pd_i}{4Et}(1 - \nu)V + \frac{pV}{K} \qquad (6\text{-}27)$$

6-4-6 Cylindrical Vessel with Hemispherical Ends

One of the most commonly used configurations of pressure vessels is a cylindrical vessel with hemispherical ends, as shown in Fig. 6-4. The wall thickness of the cylindrical and hemispherical parts may be different. This is often necessary because the hoop stress in the cylinder is twice that in a sphere of the same inside diameter and wall thickness. The internal diameters of both parts are generally considered equal. In order that there should be no distortion or mismatch of hoop stress at the junction, the hoop stresses for the cylindrical part and the hemispherical part must be equal at the end junctions. Therefore,

$$\frac{pd_i}{4Et_c}[2 - \nu] = \frac{pd_i}{4Et_s}[1 - \nu] \qquad (6\text{-}28)$$

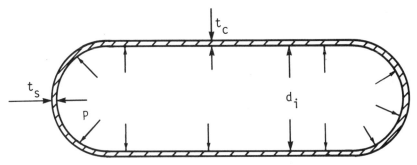

FIG. 6-4   Thin cylindrical shell with hemispherical ends.

where t_c = thickness of the cylinder, and t_s = thickness of the hemisphere. Simplification of Eq. (6-28) gives

$$\frac{t_s}{t_c} = \frac{(1 - \nu)}{(2 - \nu)} \tag{6-29}$$

6-4-7   Effects of Joints and End Plates in Vessel Fabrication

The preceding sections have assumed homogeneous materials and uniform material properties of the components. The effects of the end plates and joints will be the reduction of strength of the components due to characteristic fabrication techniques, such as riveted joints, welding, etc. To some extent, this reduction is taken into account by using a parameter, joint efficiency, in the equation of the stresses. For a thin cylindrical vessel as depicted in Fig. 6-5 the actual hoop and longitudinal stresses are given by the following equations:

$$\sigma_h = \frac{pd_i}{2t\eta_\ell} \tag{6-30}$$

$$\sigma_\ell = \frac{pd_i}{4t\eta_c} \tag{6-31}$$

where η_ℓ = efficiency of longitudinal joint, and η_c = efficiency of the circumferential joints. Similarly, for a thin sphere, hoop stress is given by

$$\sigma_h = \frac{pd_i}{4t\eta} \tag{6-32}$$

where η = the joint efficiency.

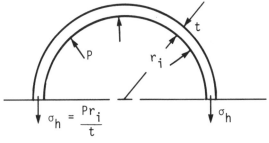

FIG. 6-5   Thin cylindrical shell subjected to internal pressure.

Notes ▪ Drawings ▪ Ideas

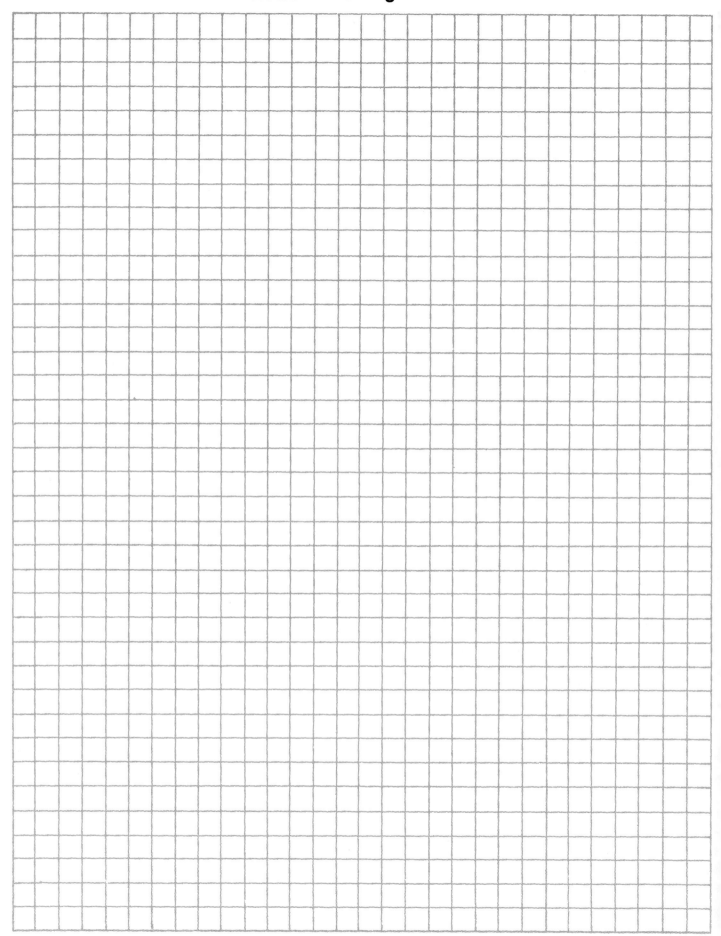

6-5 THICK CYLINDRICAL SHELLS

The theoretical treatment of thin cylindrical shells assumes that hoop stress is constant across the thickness of the shell wall and also that there is no pressure gradient across the wall. The Lamé theory for determination of the stresses in the walls of thick cylindrical shells considers a mutually perpendicular, triaxial, principal-stress system consisting of the radial, hoop or tangential, and longitudinal stresses acting at any element in the wall. For the case of the shell depicted in Fig. 6-6, subjected to internal pressure only, radial stress σ_r and hoop stress σ_h are given by

$$\sigma_r = \frac{pr_i^2}{(r_o^2 - r_i^2)} \left(\frac{r^2 - r_o^2}{r^2} \right) \tag{6-33}$$

and

$$\sigma_h = \frac{pr_i^2}{(r_o^2 - r_i^2)} \left(\frac{r^2 + r_o^2}{r^2} \right) \tag{6-34}$$

where r_i and r_o = inside and outside radii of the shell, respectively, and r = any radius such that $r_i < r < r_o$.

In order to get an expression for the longitudinal stress σ_ℓ, the shell is considered closed at both ends, as shown in Fig. 6-7. Simple consideration of the force equilibrium in the longitudinal direction yields

$$\sigma_\ell = \frac{pr_i^2}{r_o^2 - r_i^2} \tag{6-35}$$

The changes in the dimensions of the cylindrical shell can be determined by the following strain formulas:
Hoop strain:

$$\varepsilon_h = \frac{1}{E} [\sigma_h - \nu\sigma_r - \nu\sigma_\ell] \tag{6-36}$$

Longitudinal strain:

$$\varepsilon_\ell = \frac{1}{E} [\sigma_\ell - \nu\sigma_r - \nu\sigma_h] \tag{6-37}$$

It can be shown easily that diametral stress and the circumferential or hoop stress are equal. It is seen that the inside diameter-to-thickness ratio d_i/t is an important parameter in the stress formulas. For d_i/t values greater than 15, the error involved in using the thin-shell theory is within 5 percent. If, however, the mean diameter d_m

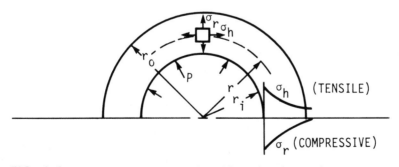

FIG. 6-6 Thick cylindrical shell subjected to internal pressure.

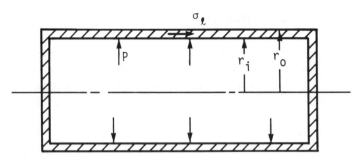

FIG. 6-7 Thick cylindrical shell closed at both ends and subjected to internal pressure.

is used for calculation of the thin-shell values instead of the inside diameter, the error reduces from 5 percent to approximately 0.25 percent at $d_m/t = 15.0$.

When the cylindrical shell is subjected to both the external pressure p_o and the internal pressure p_i, the radial and hoop stresses can be expressed by

$$\sigma_r = \frac{r_i^2 p_i - r_o^2 p_o}{r_o^2 - r_i^2} - \frac{(p_i - p_o) r_i^2 r_o^2}{r^2(r_o^2 - r_i^2)} \qquad (6\text{-}38)$$

and

$$\sigma_h = \frac{r_i^2 p_i - r_o^2 p_o}{r_o^2 - r_i^2} + \frac{(p_i - p_o) r_i^2 r_o^2}{r^2(r_o^2 - r_i^2)} \qquad (6\text{-}39)$$

It is observed from the preceding equations that the maximum value of σ_h occurs at the inner surface. The maximum radial stress σ_r is equal to the larger of the two pressures p_i and p_o. These equations are known as the *Lamé solution*. The sum of the two stresses remains constant, which indicates that the deformation of all elements in the axial direction is the same, and the cross sections of the cylinder remain plane after deformation.

The maximum shearing stress at any point in the wall of a cylinder is equal to one-half the algebraic difference between the maximum and minimum principal stresses at that point. The axial or longitudinal stress is usually small compared with the radial and tangential stresses. The shear stress τ at any radial location in the wall is given by

$$\tau = \frac{\sigma_h - \sigma_r}{2}$$

or

$$\tau = \frac{(p_i - p_o) r_i^2 r_o^2}{(r_o^2 - r_i^2) r^2} \qquad (6\text{-}40)$$

6-5-1 Compound Cylinders

Equation (6-34) indicates that there is a large variation in hoop stress across the wall of a cylindrical shell subjected to internal pressure p_i. In order to obtain a more uniform hoop-stress distribution, cylindrical shells are often fabricated by shrinking one onto the outside of another. When the outer cylinder contracts on cooling, the inner tube is brought into a state of compression. The outer cylinder will be in tension. If the compound cylinder is subjected to an internal pressure, the resultant hoop stresses will be the algebraic sum of those resulting from the internal pressure and those resulting from the shrinkage. The net result is a small total variation in hoop stress across the wall thickness. A similar effect is realized when a cylinder is wound with wire or steel tape under tension. Comings [6-4] gives a complete analysis for such systems. In order to obtain a favorable stress pattern, an analogous technique can be used by applying a sufficiently high internal pressure to produce plastic deformation in the inner part of the cylinder. On removal of this internal pressure, the residual stress persists, with the net effect of compression in the inner part and tension in the outer part of the cylinder. This process is called *autofrettage*. Harvey [6-5] discusses stress analysis for autofrettage in detail.

6-6 THERMAL STRESSES IN CYLINDRICAL SHELLS

Thermal stresses in the structure are caused by the temperature gradient in the wall and accompanying dimensional changes. In order to develop the thermal stresses, the structural member must be restrained in some way. The constraints in the thermal-stress problems in the design of vessels are usually divided into external and internal constraints. Whenever there is a significant temperature gradient across the vessel wall, thermal expansion takes place. Generally, the effects of the externally applied loads and the effects of thermal expansion or contraction are independently analyzed, and the final effects of the total combined stresses are considered additive when comparisons are made with maximum allowable stresses. It is usually assumed that thermal stresses are within the elastic range on the stress-strain curve of the vessel material. If the temperature is very high, the creep may become significant and must be considered. The induced thermal stresses often exceed the yield strength of the vessel material. Since thermal stresses are self-limiting, local plastic relaxation will tend to reduce the acting load.

The basic equations for thermal stresses can be developed by considering that a body is composed of unit cubes of uniform average temperature T. If the temperature of the unit cube is changed to T_1 such that $T_1 > T$ and is restricted, then there are three distinct cases according to the nature of restriction. If a cartesian coordinate system x, y, z is considered, then restrictions can be imagined in (1) the x direction, (2) in the x and y directions, and (3) in all three directions, i.e., in the x, y, and z directions. The corresponding thermal stresses are given by the following expressions:

$$\sigma_x = -\alpha E(T_1 - T) \tag{6-41}$$

$$\sigma_x = \sigma_y = \frac{-\alpha E}{(1 - \nu)}(T_1 - T) \tag{6-42}$$

$$\sigma_x = \sigma_y = \sigma_z = \frac{-\alpha E}{(1 - \nu)}(T_1 - T) \tag{6-43}$$

where α = coefficient of thermal expansion
  ν = Poisson's ratio
  E = modulus of elasticity

The equations are basic for direct thermal stresses under external or internal constraints and give the maximum thermal stress for the specific constraint.

The internal constraint is due to nonuniform temperature distribution in the body of the structure such that it does not allow free expansion of the individual body elements according to the local temperatures. In such a case, stresses are induced in

Notes ▪ Drawings ▪ Ideas

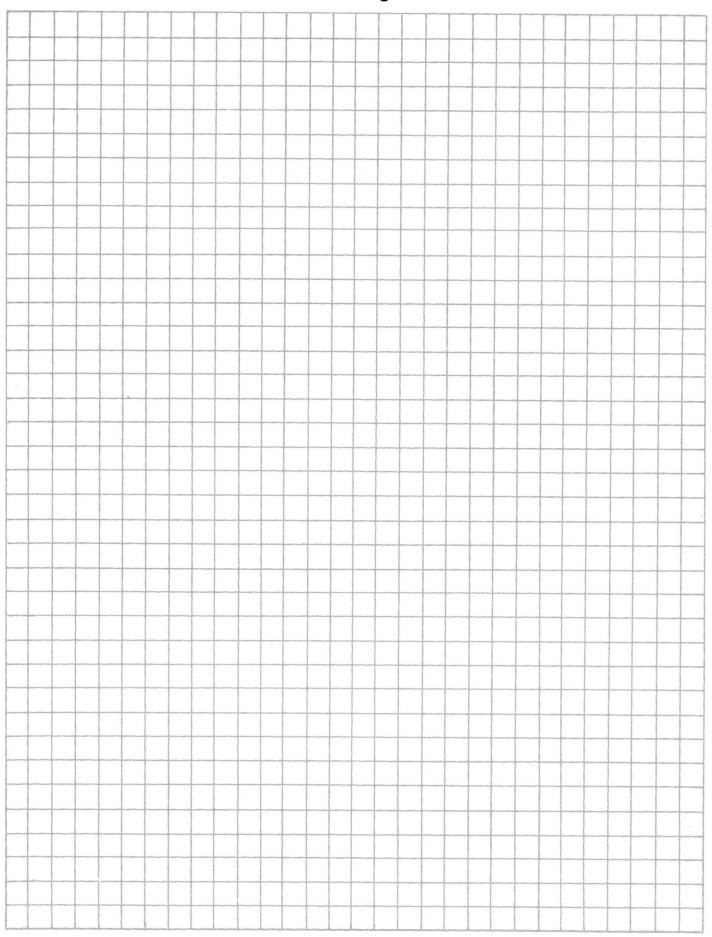

the body in the absence of any external constraints. A thick cylindrical shell may be assumed to consist of thin, mutually connected, concentric cylindrical shells. Whenever a temperature gradient exists in the vessel wall due to heat transfer, the cylindrical elements will be at different average temperatures and will expand at different rates. The individual cylindrical elements will be constrained by each other, and consequently, thermal stresses will be induced in the otherwise nonrestrained cylinder wall. Detailed treatments of the general analytical methods for solving thermal stresses caused by internal constraints are given by Gatewood [6-6], Bergen [6-7], and Goodier [6-8].

Most cylindrical-shell and pressure-vessel problems can be reduced to two-dimensional stress problems. Generally, analytical solutions are possible for relatively simpler cases.

6-6-1 Thermal Stresses in Thin Cylindrical Vessels

Consider a hollow, thin cylindrical vessel (Fig. 6-8) subjected to a linear radial temperature gradient. The temperature at the inner wall is T_1 and is greater than temperature T_2 at the outer wall. If the vessel is long and restrained at the ends, then the longitudinal stress σ_z is given by

$$\sigma_z = \frac{E\alpha(T_1 - T_2)}{2(1 - \nu)} \tag{6-44}$$

where α = coefficient of thermal expansion, and ν = Poisson's ratio.

Estimation of thermal stresses in composite cylinders is quite involved. Faupel and Fisher [6-10] consider the thermal stresses in a multishell laminate of different materials with thermal gradients through each shell.

6-6-2 Thermal Stresses in Thick Cylindrical Vessels

When a thick-walled cylindrical vessel is subjected to a thermal gradient, nonuniform deformations are induced, and consequently, thermal stresses are developed.

Fluids under high pressure and temperature are generally transported through structures such as boilers, piping, heat exchangers, and other pressure vessels. Owing to the presence of a large temperature gradient between the inner and outer walls, thermal stresses are produced in these structures. Radial, hoop, and longitudinal stresses in thick, hollow cylinders with a thermal gradient across the wall may be estimated analytically, but generally the computations are lengthy, tedious, and time-consuming. Bhaduri [6-9] developed a set of dimensionless graphs from computer solutions, and they can be used to find the thermal-stress components with a few simple calculations.

In this technique, the temperature of the shell's inner surface at radius r_i and outer surface at radius r_o are considered to

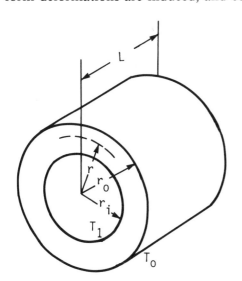

FIG. 6-8 Hollow cylinder subjected to a temperature gradient.

be T_i and T_o, respectively. The ends of the cylindrical shell are considered unrestrained. The longitudinal strain developed as a result of the thermal stresses is assumed to be uniform and constant. The temperature distribution in the shell wall is given by

$$T = T_i \left(\frac{\ln r_o/r}{\ln r_o/r_i} \right)$$

when the outer surface temperature $T_o = 0$, that is, when the temperature differences are measured relative to the outer surface temperature. The dimensionless hoop-stress function F_h, radial stress function F_r, and longitudinal stress function F_z are obtained analytically for cylindrical shells of various thickness ratios. The stress functions are given by the following equations:

$$F_h = \frac{2(1-\nu)\sigma_h}{\alpha E T_i} = \frac{1}{\ln R_o} \left[1 - \ln \frac{R_o}{R} - \frac{(R_o/R)^2 + 1}{R_o^2 - 1} \ln R_o \right]$$

$$F_r = \frac{2(1-\nu)\sigma_r}{\alpha E T_i} = \frac{1}{\ln R_o} \left[\ln \frac{R}{R_o} + \frac{(R_o/R)^2 - 1}{R_o^2 - 1} \ln R_o \right] \qquad (6\text{-}45)$$

$$F_z = \frac{2(1-\nu)\sigma_z}{\alpha E T_i} = \frac{1}{\ln R_o} \left(1 - 2 \ln \frac{R_o}{R} - \frac{2}{R_o^2 - 1} \ln R_o \right)$$

where σ_h = hoop stress
$\qquad \sigma_r$ = radial stress
$\qquad \sigma_z$ = longitudinal stress
$\qquad \nu$ = Poisson's ratio
$\qquad \alpha$ = coefficient of thermal expansion
$\qquad E$ = Young's modulus
$\qquad R = r/r_i$
$\qquad R_o = r_o/r_i$

The stress functions are shown graphically for the radii ratios $r_o/r_i = 2.0$, 2.5, and 3.00, respectively in Fig. 6-9. These curves are general enough to compute the hoop, radial, and shear stresses produced by temperature gradients encountered in most cylindrical-shell designs.

The procedure is simple. The value of R_o must be known to select the appropriate curve. For a particular value of R, the corresponding values of the stress functions can be read from the ordinate. The stresses at any radial location of the shell wall can be easily calculated from the known values of the stress function, property values of the shell material, and temperature of the inside surface of the shell.

6-7 FABRICATION METHODS AND MATERIALS

Welding is the most common method of fabrication of pressure-vessel shells. Structural parts such as stiffening rings, lifting lugs, support clips for piping, internal trays, and other parts are also attached to the vessel wall by welding. Welded joints are used for pipe-to-vessel connections to ensure optimum leak-proof design, particularly when the vessel contains hazardous fluid at a very high temperature. A structure whose parts are joined by welding is called a *weldment*. The most widely used industrial welding method is arc welding. It is any of several fusion welding processes wherein the heat of fusion is generated by electric arc. Residual stresses in a weld

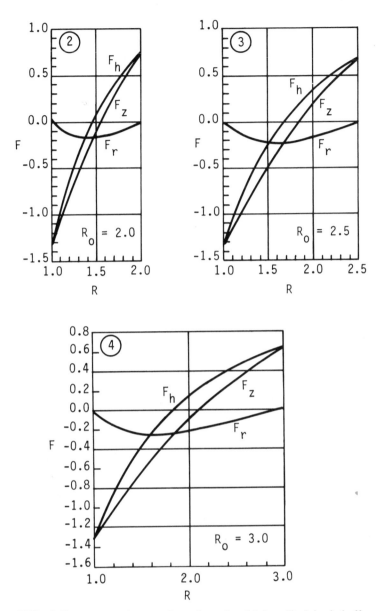

FIG. 6-9 Thermal stress functions for thick cylindrical shell. *(From Ref. [7-9].)*

and in the adjoining areas cannot be avoided. They are quite complex. If the weld residual stress is superposed on the stress due to external loads and the resultant stress exceeds the yield point of the material, local plastic yielding will result in redistribution of the stress in ductile materials. A good weld requires a highly ductile material. In order to prevent loss of ductility in the welded region, low-carbon steels with less than 0.35 percent carbon content are used as construction materials. Carbon itself is a steel hardener. However, in the presence of a manganese content of 0.30 to 0.80 percent, carbon does not cause difficulties when present in steel up to 0.30 percent. Welding is a highly specialized manufacturing process. In pressure-vessel fabrication, the designer has to follow the code developed by the American Soci-

Notes ▪ Drawings ▪ Ideas

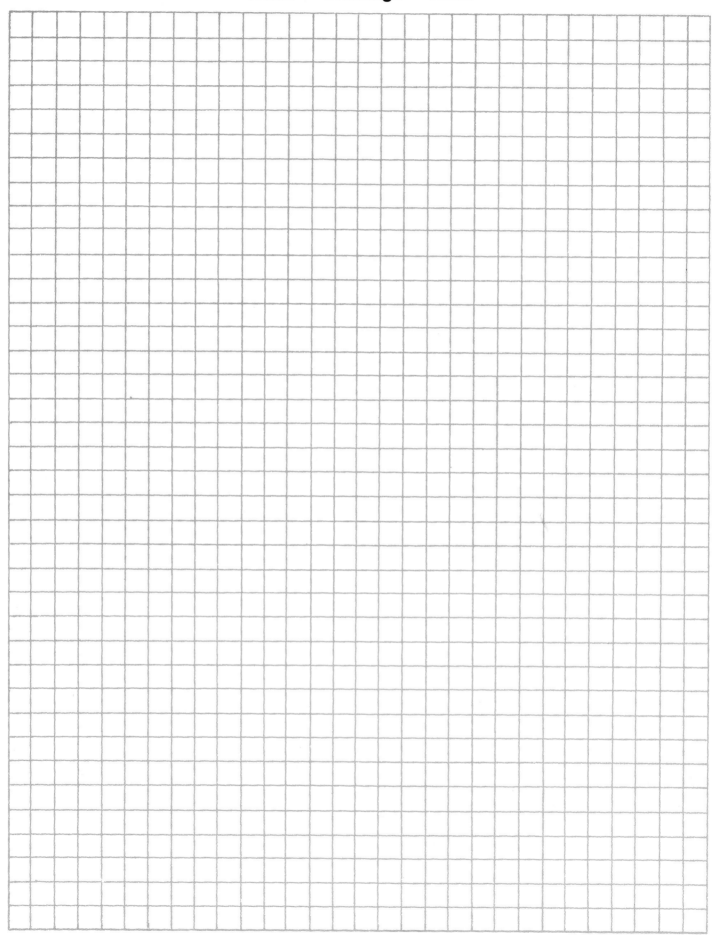

ety of Mechanical Engineers. Chuse [6-3] provides simplified guidelines for designing welding joints for pressure vessels. For design purposes, welding is classified into three basic types: groove, fillet, and plug welds. Welding joints are described by the position of the pieces to be joined and are divided into five basic types: butt, tee, lap, corner, and edge. Bednar [6-2] describes various types of welds and outlines methods for stress calculation.

6-7-1 Construction Materials

A designer of a cylindrical pressure vessel should be familiar with commonly used construction materials to be able to specify them correctly in the design and in material specification. The selection of materials for code pressure vessels has to be made from the code-approved material specifications.

6-8 DESIGN OF PRESSURE CYLINDERS

Cylindrical shells are commonly used in industrial applications for their adequate structural strength, ease of fabrication, and economical consumption of material. The vessels, depending on their function, can be vertical or horizontal. A vertical orientation is often preferred for thin-walled vessels operating under low internal pressure because the additional bending stresses resulting from the weights of the vessel itself and the fluid in the vessel.

Cylindrical shells of such elastic materials as steels, nonferrous metals, and most alloys that operate under internal pressures up to 10 MPa are fabricated mostly of rolled and subsequently welded sheets. The joint connections of cylindrical shells made of copper or brass sheets are made by soldering with suitable solders. Cylindrical vessels of steel operating under pressures greater than 10 MPa are commonly fabricated of forged pieces that are heat-treated. Cylindrical shells of brittle materials for vessels operating under low internal pressures (approximately up to 0.8 MPa) are molded. Cast shells are usually fabricated with a bottom. In some cases, cast shells are made of elastic metals and their alloys.

The fabrication of the cylindrical vessels by the rolling of sheets is a very common method of manufacturing low- and medium-pressure (1.75 to 10.0 MPa) vessels.

Azbel and Cheremisinoff [6-1] give the following general guidelines for designing welded and soldered cylindrical shells:

1. The length of the seams should be minimized.
2. The minimum number of longitudinal seams should be provided.
3. Longitudinal and circumferential seams must be butt welded.
4. All joints should be accessible for inspection and repair.
5. It is not permissible to provide holes, access holes, or any opening on the seams, especially on longitudinal seams.

Shell thicknesses of vessels operating under very low pressures are not designed; they are selected on the basis of manufacturing considerations. Durability is estimated from the available information regarding corrosion resistance of the material. The minimum thicknesses of a shell rolled from sheet metals are given as follows: carbon and low-alloy steel, 4 mm; austenite steel, 3.0 to 4.0 mm; copper, 2.5 to 3.0 mm; and cast materials, 2.0 to 2.5 mm.

6-8-1 Design of Welded Cylindrical Shells

The structural design of cylindrical shells is based on membrane theory. In order to express the design relations in convenient form, it is necessary to (1) select the strength theory best reflecting the material behavior, (2) consider the weakening of the construction induced by welding and other connections, (3) consider wall thinning as a result of corrosion effects, and (4) establish a factor of safety and allowable stresses.

The distortion-energy theory gives the basic design stress as

$$\sigma_d = [\sigma_t^2 + \sigma_\ell^2 + \sigma_r^2 - 2(\sigma_t\sigma_\ell + \sigma_t\sigma_r + \sigma_\ell\sigma_r)]^{1/2} \tag{6-46}$$

where σ_t, σ_ℓ, and σ_r = principal circumferential, longitudinal, and radial stresses, respectively. For a thin-walled cylindrical shell operating under an internal pressure p_i, the radial stress is assumed to be equal to zero, and the longitudinal stress induced in the shell is given by

$$\sigma_\ell = \frac{p_i d_i}{4t} \tag{6-16}$$

where d_i = inside diameter, and t = the thickness of the shell. The circumferential stress induced in the shell is given by

$$\sigma_t = \frac{p_i d_i}{2t} \tag{6-12}$$

When the deformation reaches plastic state, a definite amount of deformation energy is assumed (see Ref. [6-1], and the value of the Poisson's coefficient v is taken as equal to 0.5.

Substituting Eqs. (6-16) and (6-12) into Eq. (6-46) and assuming the value of Poisson's coefficient $v = 0.5$, the design stress can be expressed as

$$\sigma_d = 0.87 \frac{p_i r_o}{t} \tag{6-47}$$

where r_o = external radius of the shell. The external radius $r_o = d_o/2$, where d_o = external diameter of the shell.

It is observed that according to the distortion-energy theory, considering the combined effect of the longitudinal and tangential stresses, the design stress for plastic material is 13.0 percent less compared with the maximum value of the main stress. The allowable stress σ_a is therefore given by

$$\sigma_a = \frac{p_i d_o}{2.3t} \tag{6-48}$$

And the thickness of the shell t is given by

$$t = \frac{p d_o}{2.3\sigma_a} \tag{6-49}$$

Introducing the joint efficiency factor η_j and allowance factor t_c for corrosion, erosion, and negative tolerance of the shell thickness, the following relation for design thickness is obtained:

$$t_d = \frac{p d_o}{2.3\sigma_a \eta_j} + t_c \tag{6-50}$$

Azbel and Cheremisinoff [6-1], on the basis of similar analysis, give the following design formula for the determination of shell thickness:

$$t_d = \frac{p_i d_i}{2.3\eta_f\sigma_a - p_i} + t_c \tag{6-51}$$

where $d_o/d_i \leq 1.5$.

The proper selection of allowable stress to provide safe operation of a vessel is an important design consideration. The allowable stress value is influenced by a number of factors, such as (1) the strength and ductility of the material, (2) variations in load over time, (3) variations in temperature and their influence on ductility and strength of the material, and (4) the effects of local stress concentration, impact loading, fatigue, and corrosion.

In shell design, the criterion of determining the allowable stresses in an environmentally moderate temperature is the *ultimate tensile strength*. The pressure-vessel codes [6-3] use a safety factor of 4 based on the yield strength S_y for determining the allowable stresses for pressure vessels. The safety factor η_y is defined as

$$\eta_y = \frac{S_y}{\sigma_a} \tag{6-52}$$

It is known that for ductile metals, an increase in temperature results in an increase in ductility and a decrease in the yield-strength value. The allowable stress at higher temperatures can be estimated by using the following relation:

$$\sigma_a = \frac{S_y^T}{\eta_y} \tag{6-53}$$

where S_y^T = yield strength of the material at a given operating temperature.

Implied in the preceding analysis is an assumption that no creeping of the shell material is involved. This is generally a valid assumption for ferrous metals under load at temperatures less than 360°C. However, at higher operating temperatures, the material creeps under load, and an increase in stress takes place with time. The resulting stresses do not exceed the yield-strength values. It is to be noted that the yield point at high temperatures cannot generally be used as a criterion for allowance of the stressed state because creep of the material may induce failures as a result of the increased deformation over a long period of time. The basic design criterion for shells operating at moderate temperatures is the stability of size and dimensional integrity of the loaded elements. For a shell operating at higher temperatures, increase in size, which ensures that creep does not exceed a tolerable limit, should be considered. When a loaded material is under creep, there is an associated relaxation of stress. The stress decreases over time as a result of plastic deformation. The creep rate is dependent on temperature and state of stress in the metal. Azbel and Cheremisinoff [6-1] discuss the effects of creep and stress relaxation on loaded shells and give a formula for determining the allowable stresses and safety factors on the basis of fatigue stress.

REFERENCES

6-1 D. S. Azbel and N. P. Cheremisinoff, *Chemical and Process Equipment Design—Vessel Design and Selection,* Ann Arbor Science Publishers, Ann Arbor, Michigan, 1982.

Notes ▪ Drawings ▪ Ideas

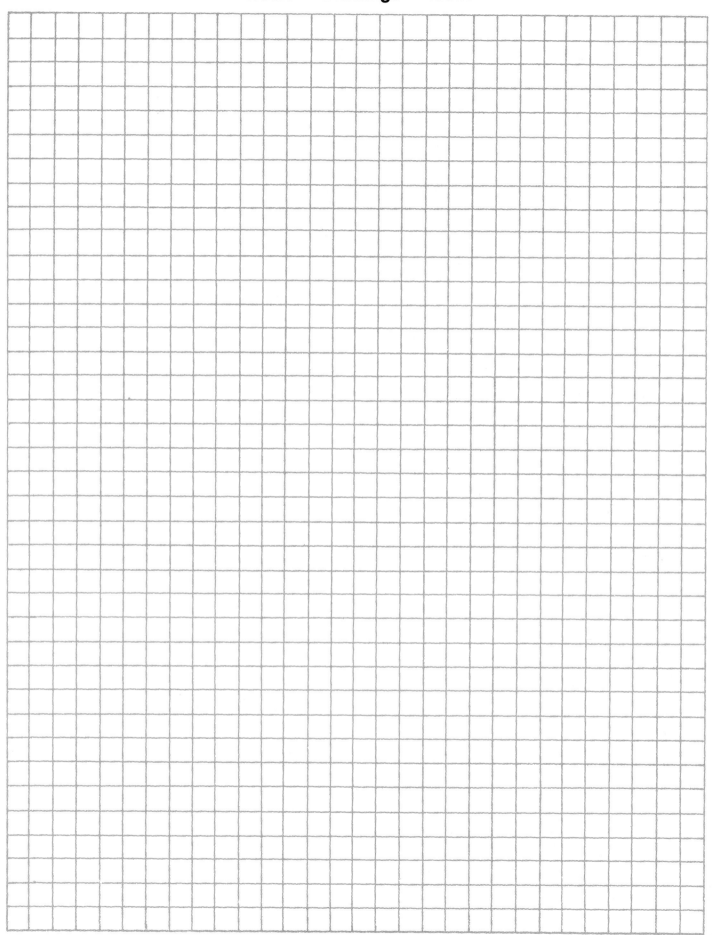

6-2   H. H. Bednar, *Pressure Vessel Design Handbook,* Van Nostrand Reinhold, New York, 1981.

6-3   R. Chuse, *Pressure Vessel—The ASME Code Simplified,* 5th Ed., McGraw-Hill, New York, 1977.

6-4   E. W. Comings, *High Pressure Technology,* McGraw-Hill, New York, 1956.

6-5   J. F. Harvey, *Pressure Vessel Design: Nuclear and Chemical Applications,* D. Van Nostrand, Princeton, N.J., 1963.

6-6   B. L. Gatewood, *Thermal Stresses,* McGraw-Hill, New York, 1957.

6-7   D. Bergen, *Elements of Thermal Stress Analysis,* C. P. Press, Jamaica, N.Y., 1971.

6-8   J. N. Goodier, ''Thermal Stress in Pressure Vessel and Piping Design, Collected Papers,'' *ASME,* New York, 1960.

6-9   S. Bhaduri, ''Thermal Stresses in Cylindrical Sheels,'' *Machine Design,* vol. 52, no. 7, April 10, 1980.

6-10 J. H. Faupel and F. E. Fisher, *Engineering Design,* Wiley, New York, 1981.

Notes ▪ Drawings ▪ Ideas

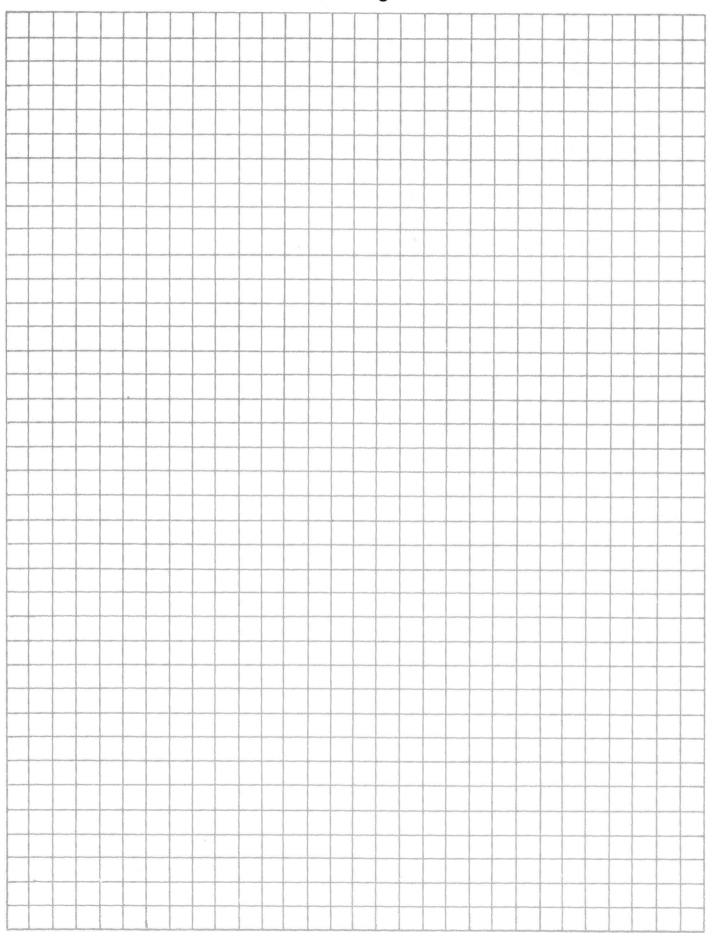

chapter 7
TOLERANCES

CHARLES R. MISCHKE, Ph.D., P.E.

Professor of Mechanical Engineering
Iowa State University
Ames, Iowa

Notes ▪ Drawings ▪ Ideas

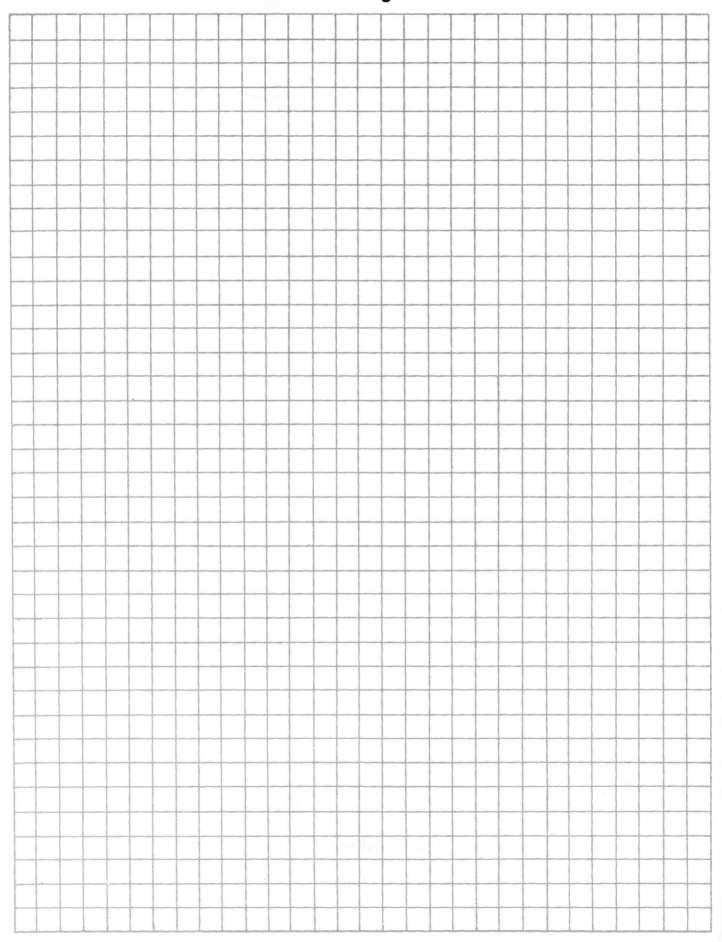

7-1 TOLERANCES†

When an aggregate of several parts is assembled, the gap, grip, or interference is related to dimensions and tolerances of the individual parts. Consider an array of parallel vectors as depicted in Fig. 7-1, the x's directed to the right and the y's directed to the left. They may be treated as scalars and represented algebraically. Let t_i be the bilateral tolerance on $\bar{x}_i$ and t_j be the bilateral tolerance on $\bar{y}_j$, all being positive numbers. The gap remaining short of closure is called z and may be viewed as the slack variable permitting summation to zero. Thus,

$$(x_1 + x_3 + \cdots) - (y_2 + y_4 + \cdots) - z = 0$$

or

$$z = \Sigma x_i - \Sigma y_j \tag{7-1}$$

The largest gap z exists when the right-tending vectors are the largest possible and the left-tending vectors are the smallest possible. Expressing Eq. (7-1) in terms of the greatest deviations from the means gives

$$z_{\max} = \Sigma(\bar{x}_i + t_i) - \Sigma(\bar{y}_j - t_j) = \Sigma\bar{x}_i - \Sigma\bar{y}_j + \sum_{\text{all}} t$$

Similarly, for the smallest gap

$$z_{\min} = \Sigma(\bar{x}_i - t_i) - \Sigma(\bar{y}_j + t_j) = \Sigma\bar{x}_i - \Sigma\bar{y}_j - \sum_{\text{all}} t$$

The mean of z is

$$\bar{z} = \tfrac{1}{2}(z_{\max} + z_{\min}) = \tfrac{1}{2}\left[(\Sigma\bar{x}_i - \Sigma\bar{y}_j + \Sigma t) + (\Sigma\bar{x}_i - \Sigma\bar{y}_j - \Sigma t)\right]$$
$$\bar{z} = \Sigma\bar{x}_i - \Sigma\bar{y}_j \tag{7-2}$$

The bilateral tolerance of z is

$$t_z = \tfrac{1}{2}(z_{\max} - z_{\min}) = \tfrac{1}{2}\left[(\Sigma\bar{x}_i - \Sigma\bar{y}_j + \Sigma t) - (\Sigma\bar{x}_i - \Sigma\bar{y}_j - \Sigma t)\right]$$
$$t_z = \sum_{\text{all}} t \tag{7-3}$$

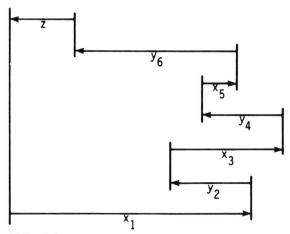

FIG. 7-1 An array of parallel vectors.

†See Refs. [7-1] and [7-2].

Notes ▪ Drawings ▪ Ideas

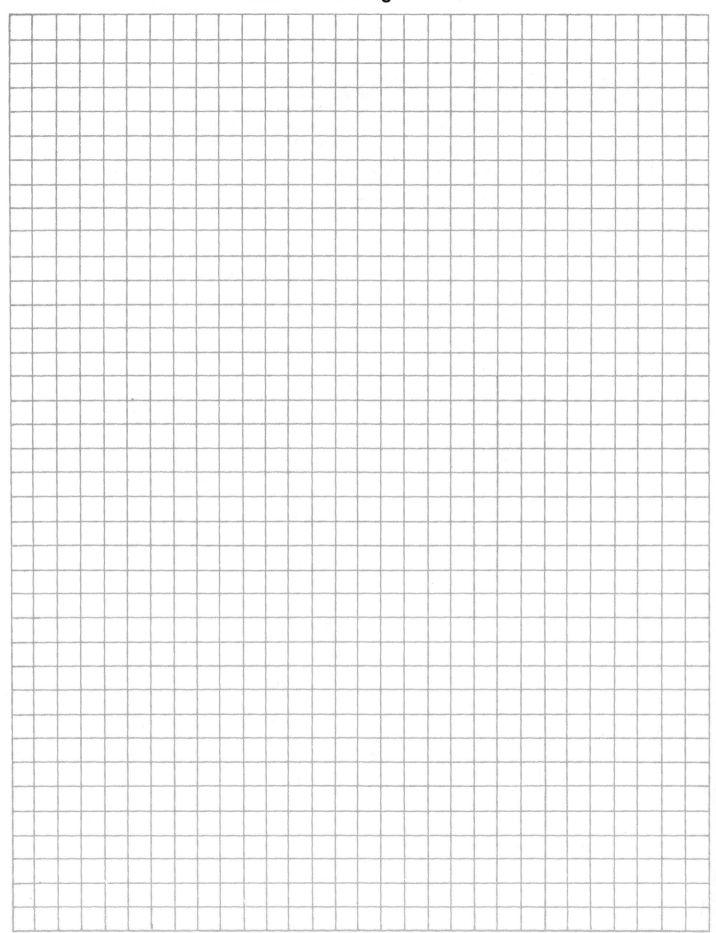

Equation (7-3) gives rise to expressions such as "the stacking of tolerances" in representing the condition at the gap. The bilateral tolerance at the gap is the sum of all bilateral tolerances of vectors in the chain, regardless of their orientation (left- or right-tending). If the gap is an interference, then z is a right-tending vector. For all instances to be interference fits, both z_{max} and z_{min} have to be negative.

The vectors can represent the thickness of material bodies assembled and abutting in various ways. The vector z can represent a clearance or interference which is to be maintained.

EXAMPLE 1. A pin and a hollow cap are assembled as shown in Fig. 7-2 and the assembly results in a gap. What are the largest and smallest values of the gap? We write

$$\bar{z} = \Sigma\bar{x}_i - \Sigma\bar{y}_j = 1.000 - 0.995 = 0.005 \text{ in}$$

$$t_z = \sum_{\text{all}} t = 0.001 + 0.002 = 0.003 \text{ in}$$

$$z_{max} = \bar{z} + t_z = 0.005 + 0.003 = 0.008 \text{ in}$$

$$z_{min} = \bar{z} - t_z = 0.005 - 0.003 = 0.002 \text{ in}$$

If the problem is inverted, that is, the gap must be $z = 0.006 \pm 0.004$ in and $y_1 = 0.995 \pm 0.002$ in, what is the dimension x and its tolerance?

$$t_z = \sum_{\text{all}} t = t_1 + t_2 = t_1 + 0.002 = 0.004 \text{ in} \qquad \therefore \quad t_1 = 0.002 \text{ in}$$

$$\bar{z} = \Sigma\bar{x}_i - \Sigma\bar{y}_j = \bar{x}_1 - 0.995 = 0.006 \text{ in} \qquad \therefore \quad \bar{x}_1 = 1.001 \text{ in}$$

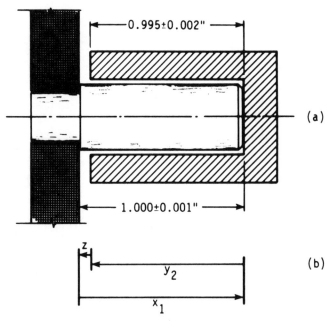

FIG. 7-2 (*a*) A pin-and-cap assembly and associated gap; (*b*) parallel vectors allowing description of gap.

Notes · Drawings · Ideas

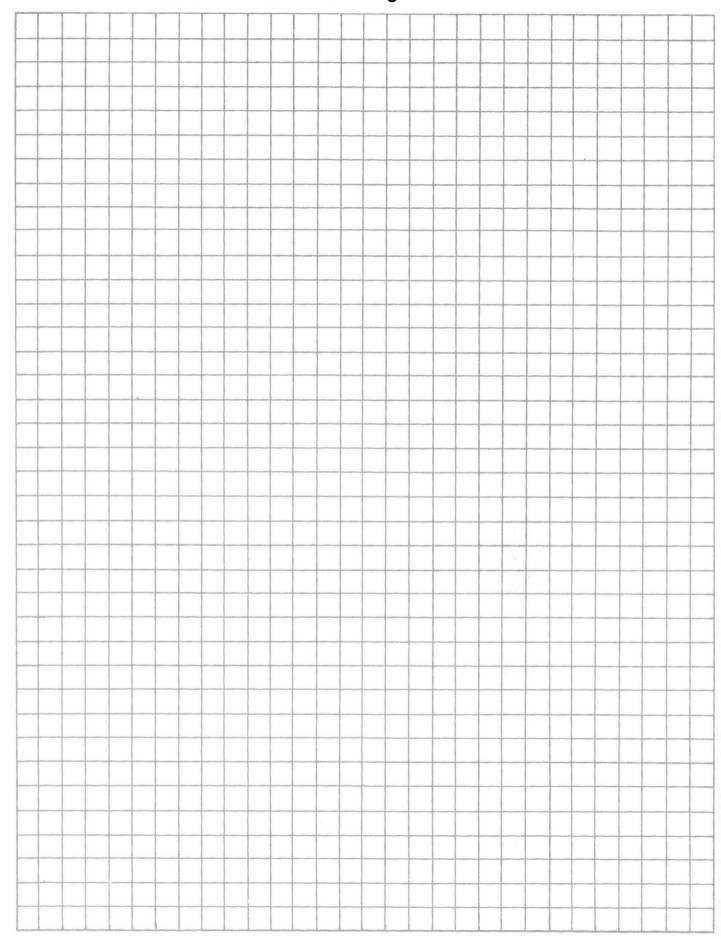

Continuing the inversion with the gap required, $z = 0.010 \pm 0.001$ in, then. Thus

$$t_z = \sum_{\text{all}} t = t_1 + t_2 = t_1 + 0.002 = 0.001 \text{ in} \qquad \therefore \quad t_1 = -0.001 \text{ in}$$

This is impossible through control of the dimension x_1 and its tolerance. If $t_z = 0.001 = \sum_{\text{all}} t$, the tolerance on the dowel pin and the cap hole must be controlled more closely. Until methods of manufacture are reconsidered such that tolerances on the order of 0.0005 in are available, no decision will satisfy the gap requirement of $z = 0.010 \pm 0.001$ in.

EXAMPLE 2. In the pin-washer-snap-ring assembly depicted in Fig. 7-3 describe the dimensional range on the gap given that

$$x_1 = 1.385 \pm 0.006 \text{ in} \qquad y_2 = 0.125 \pm 0.001 \text{ in}$$
$$y_3 = 1.000 \pm 0.002 \text{ in} \qquad y_4 = 0.250 \pm 0.001 \text{ in}$$

Now

$$\bar{z} = \Sigma \bar{x}_i - \Sigma \bar{y}_j = 1.385 - 0.125 - 1.000 - 0.250 = 0.010 \text{ in}$$

$$t_z = \sum_{\text{all}} t = 0.006 + 0.001 + 0.002 + 0.001 = 0.010 \text{ in.}$$

$$z_{\max} = \bar{z} + t_z = 0.010 + 0.010 = 0.020 \text{ in}$$

$$z_{\min} = \bar{z} - t_z = 0.010 - 0.010 = 0 \text{ in}$$

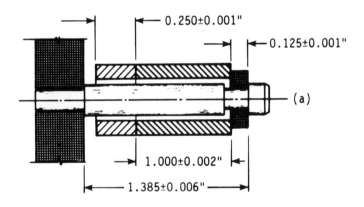

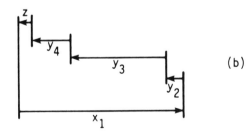

FIG. 7-3 (*a*) A pin-washer-snap-ring assembly and associated gap; (*b*) parallel vectors describing gap.

Notes ▪ Drawings ▪ Ideas

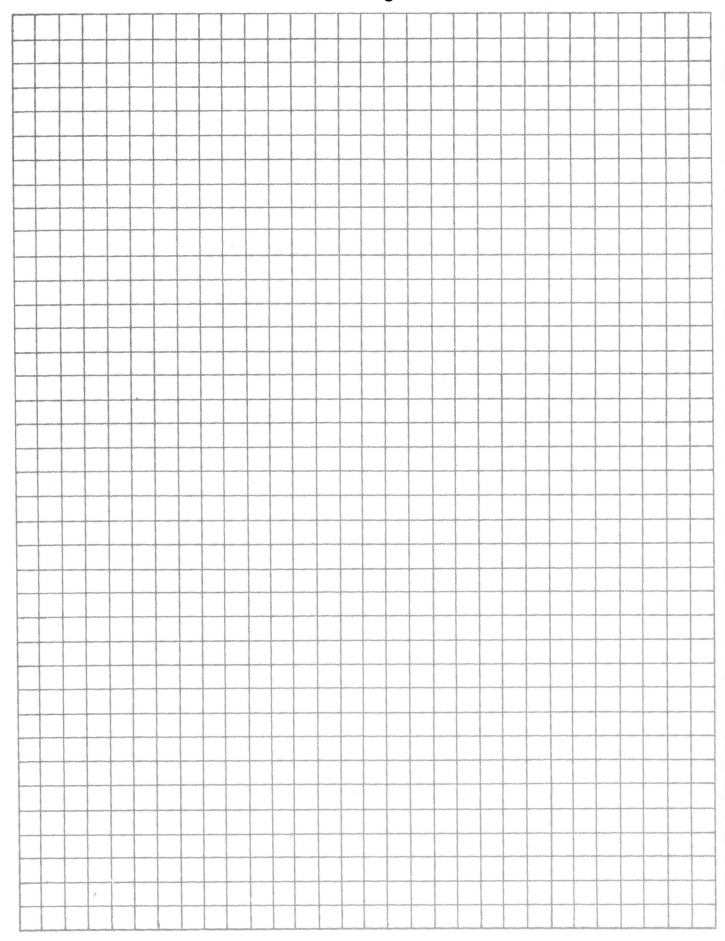

Table 7-1 depicts a tabular form useful in predicting $\bar{z}$ and t_z. Invert the problem. Given that the washer and the snap-ring are commercial parts purchased from vendors, how do we control the plate and pin length to achieve $z = 0.005 \pm 0.004$ in? Equation (7-1) requires

$$\bar{z} = \Sigma\bar{x}_i - \Sigma\bar{y}_j = \bar{x}_1 - (0.125 + \bar{y}_3 + 0.250) = 0.005 \text{ in}$$

from which

$$\bar{x}_1 - \bar{y}_3 = 0.380 \text{ in}$$

and

$$t_z = 0.004 = \sum_{\text{all}} t = (t_1 + 0.001 + t_3 + 0.001)$$

from which

$$t_{x_1} + t_{y_3} = 0.002$$

If $\bar{y}_3$ must be 1 in, then

$$\bar{x}_1 = 0.380 + \bar{y}_3 = 0.380 + 1.000 = 1.380 \text{ in}$$

The tolerances must be apportioned between x_1 and y_3 such that the sum is 0.002 in and they can be held by available manufacturing processes. If each can be 0.001 in, then

$$x_1 = 1.380 \pm 0.001 \text{ in}$$
$$y_3 = 1.000 \pm 0.001 \text{ in}$$

EXAMPLE 3. Do the specifications displayed in Fig. 7-4 produce an interference fit? These specifications give

$$z_{\text{max}} = \Sigma\bar{x}_i - \Sigma\bar{y}_j + \sum_{\text{all}} t = \frac{1.001}{2} - \frac{1.000}{2} + \frac{0.001}{2} + \frac{0.001}{2} = 0.0015 \text{ in}$$

$$z_{\text{min}} = \Sigma\bar{x}_i - \Sigma\bar{y}_j - \sum_{\text{all}} t = \frac{1.001}{2} - \frac{1.000}{2} - \left(\frac{0.001}{2} + \frac{0.001}{2}\right) = -0.0005 \text{ in}$$

$$\bar{z} = \Sigma\bar{x}_i - \Sigma\bar{y}_j = \frac{1.001}{2} - \frac{1.000}{2} = 0.0005 \text{ in}$$

TABLE 7-1 Absolute Tolerance Worksheet

i	t	x_i	y_i
1	0.006	1.385	
2	0.001		0.125
3	0.002		1.000
4	0.001		0.250
	$\Sigma = 0.010$	1.385	1.375
		-1.375	
		$\bar{z} = 0.010$	

Notes · Drawings · Ideas

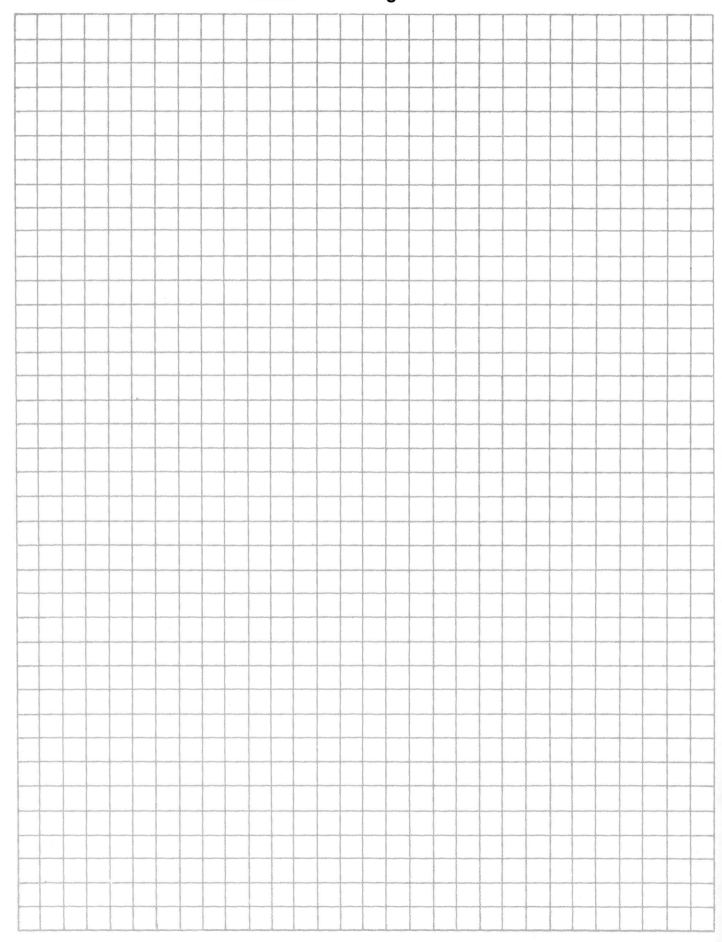

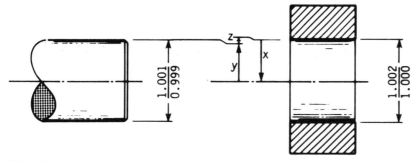

FIG. 7-4 A spindle-bushing assembly with max/min dimensioning.

Some assemblies will be interference fits. If the objective was to produce all interference fits, it has not been achieved with these specifications.

EXAMPLE 4. Figure 7-5 shows a journal-bushing assembly with unilateral tolerances. What is the description of the radial clearances which result from these specifications? With these specifications we find

$$z_{max} = \Sigma \bar{x}_i - \Sigma \bar{y}_j + \sum_{all} t = \left(\frac{B}{2} + \frac{b}{4} \right) - \left(\frac{D}{2} - \frac{d}{4} \right) + \left(\frac{b}{4} + \frac{d}{4} \right)$$

$$= \frac{B - D}{2} + \frac{b + d}{2}$$

$$z_{min} = \Sigma \bar{x}_i - \Sigma \bar{y}_j - \sum_{all} t = \left(\frac{B}{2} + \frac{b}{4} \right) - \left(\frac{D}{2} - \frac{d}{4} \right) - \left(\frac{b}{4} + \frac{d}{4} \right)$$

$$= \frac{B - D}{2}$$

$$\bar{z} = \Sigma \bar{x}_i - \Sigma \bar{y}_j = \left(\frac{B}{2} + \frac{b}{4} \right) - \left(\frac{D}{2} - \frac{d}{4} \right) = \frac{B - D}{2} + \frac{b + d}{4}$$

All these examples are describing absolute tolerancing schemes where *no* scrap would be produced if the gap were to be inspected. In this scheme, inspection of component dimensions ensures that there are no violations of the gap tolerance. Holding parts to tolerance in this fashion is sure. The observation that gap dimensions anywhere near gap limits is an exceedingly rare event leads to consideration of statistical tolerancing schemes. If occasional scrapping of an assembly is acceptable, or if occasional selective assembly can be accommodated, then much wider member tolerances are possible with attendant savings in manufacture. The basic equation for the assembly is

$$z = \Sigma \bar{x}_i - \Sigma \bar{y}_j$$

If all x's and y's are uncorrelated random variables, then

$$\mu_z = \Sigma \mu_{x_i} - \Sigma \mu_{y_j}$$

since sums (or differences) of random variables propagate their means as a sum (or difference). The variances propagate additively for both sums and differences [7-3]:

$$\sigma_z^2 = \Sigma\sigma_{x_i}^2 + \Sigma\sigma_{y_j}^2 = \sum_{\text{all}}\sigma^2$$

If bilateral tolerances are some consistent multiple of the standard deviation, such as 3, which encompasses 0.997 of the population (two-tailed), then it can be shown that

$$t_z = \sqrt{\sum_{\text{all}}t^2}$$

For Example 1,

$$\mu_z = \Sigma\mu_{x_i} - \Sigma\mu_{y_j} = 1.000 - 0.995 = 0.005 \text{ in}$$

as before, but

$$t_z = \sqrt{\sum_{\text{all}}t^2} = (0.001^2 + 0.002^2)^{1/2} = 0.002\ 236 \text{ in}$$

This is smaller than with absolute tolerances allowing t_1 and t_2 to be larger and is usually cheaper to create. The chance of exceeding $t_z = \pm 0.002\ 236$ is $1 - 0.9973 = 0.0027$, or about 3 in 1000. In Example 2,

$$\mu_z = \Sigma\mu_{x_i} - \Sigma\mu_{yj} = 1.385 - 0.125 - 1.000 - 0.250$$

$$= 0.010 \text{ in}$$

as before, but

$$t_z = \sqrt{\sum_{\text{all}}t^2} = (0.006^2 + 0.001^2 + 0.002^2 + 0.001^2)^{1/2}$$

$$= 0.006\ 481 \text{ in}$$

which is somewhat more than half the previous assignment. Table 7-2 depicts a convenient tabular form for determining μ_z and t_z. The chance of exceeding $t_z = \pm 0.006\ 481$ is $1 - 0.9973$, or 0.0027. One can loosen the tolerance on the pin to

$$t_z = 0.010 = (t_1^2 + 0.001^2 + 0.002^2 + 0.001^2)^{1/2}$$

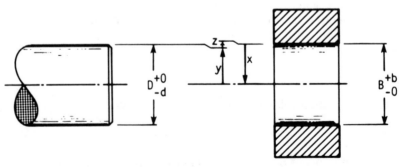

FIG. 7-5   A journal-bushing assembly with unilateral tolerances.

TABLE 7-2 Statistical Tolerance Worksheet

i	t	t^2	x_i	y_i
1	0.006	36×10^{-6}	1.385	
2	0.001	1×10^{-6}		0.125
3	0.002	4×10^{-6}		1.000
4	0.001	1×10^{-6}	___	0.250
	$\Sigma = 0.010$	42×10^{-6}	1.385	1.375
		$\sqrt{t^2} = 6.48 \times 10^{-3}$	-1.375	
			$\bar{z} = 0.010$	

from which $t_1 = 0.009\,695$. The tolerance on x_1 can be raised from ± 0.006 to ± 0.009 with only a $1 - 0.9988 = 0.0012$ chance of exceeding ± 0.010 in on the gap, or a 0.0006 chance of an interference. Appreciation of the statistical overtones of assembly raises the burdens on in-house manufacture.

REFERENCES

7-1 Joseph E. Shigley and Charles R. Mischke, *Mechanical Engineering Design,* 5th ed., McGraw-Hill, New York, 1989.

7-2 M. F. Spotts, *Dimensioning and Tolerancing For Quality Production,* Prentice-Hall, Englewood Cliffs, N.J., 1983. (Excellent bibliography on standards and handbooks, dimensioning and tolerancing, quality control, gauging and shop practice, probability and statistics.)

7-3 C. R. Mischke, *Mathematical Model Building,* 2d rev. ed., Iowa State University Press, Ames, Iowa, 1980.

INDEX

ABOUT THE EDITORS

Joseph E. Shigley has had a long and distinguished career both as an educator and as a consultant in machine design and mechanical engineering. He was a professor of mechanical engineering at the University of Michigan for 21 years. He is a Fellow of the American Society of Mechanical Engineers. He received the ASME Mechanisms Committee Award in 1974, the Worchester Reed Warner Medal in 1977, and the Machine Design Award in 1985. He is the author or co-author of eight McGraw-Hill books.

Charles R. Mischke is Professor of Mechanical Engineering at Iowa State University, a consultant to industry, contributor of many technical papers, author or co-author of five books, and co-editor of the *Standard Handbook of Machine Design*. He received the Ralph Teeter Award of the Society of Automotive Engineers in 1977, and the University's Outstanding Teacher Award in 1980. A Fellow of the American Society of Mechanical Engineers (ASME), he also serves on its Reliability, Stress Analysis, and Failure Prevention Executive Committee.